"Barry Glassner's *Gospel of Food* is a well-timed brick through the plate-glass window of conventional wisdom, pretentiousness, willful ignorance, and political correctness through which most writers on food look out at the world. He gives you plenty to think about."   —Anthony Bourdain

"In his previous book, the sociologist Barry Glassner showed us why the only thing we have to fear is fear itself. In this latest effort, he applies the same pragmatism to the ever-more-complex interlocking worlds of food, gastronomy, and health. The result is another work of refreshingly sensible skepticism about all manner of nostrums that offers some sound advice to avoid extremes, in either practice or expectation. Along the way, Glassner gives sharp views of the restaurant scene and piquant portraits of some of its stars."   —*Atlantic Monthly*

"Mr. Glassner's latest book, *The Gospel of Food*, posits that misguided (and misguiding) gospelmongers are persuading Americans to eat according to ever-more irrational notions and sillier motivations. It's plain that part of his own motivation is a genuine love of food and indignation at seeing a great common blessing made the sparring ground of officious sectarians who frequently don't know what they're talking about. . . . As in *The Culture of Fear*, Mr. Glassner exposes the strained interpretations, 'prejudices dressed up as science,' and pure fabrications behind much-received wisdom simply by checking out sources with an eye to original meanings . . . [and] he gets to the heart of something deeply wrong on the national food scene: Most of us want the right food to make us healthy and wise as individuals and as a society, without trusting ourselves to know rightness."   —*New York Times*

"*The Gospel of Food* is pure fun to read. In pitting pundits against each other— the *New York Times*' Jane Brody versus the former *Los Angeles Times* writer Emily Green, the *New York Times Magazine*'s Gary Taubes versus the *Washington Post*'s Sally Squires—and citing numerous food writers, including M. F. K. Fisher, Alice Waters, Ruth Reichl, and Jeffrey Steingarten, the book is a deliciously gossipy, delightfully acerbic, voyeuristic foray into the inner circle of the culinary cognoscenti. . . . Glassner is methodical and relentless in his exploration, fierce in his finger-pointing . . . and genuinely concerned about our growing disassociation with, and emotional baggage around, food."
                    —*Los Angeles Times Book Review*

"In his latest debunking project (after *The Culture of Fear*), sociologist Glassner argues that 'everything you think you know about food is wrong.' And Glassner really does take on almost everything, from Atkins to vegans, with particularly hard jabs at those who, in the name of nutrition, take the fun out of food. . . . Only two conventional bits of wisdom survive Glassner's skeptical approach: The rich really are thinner than the poor and four-star restaurant cooking really is delicious."   —*Publishers Weekly*

"While Glassner examines nearly every issue populating the food landscape, *The Gospel of Food* shines brightest when he turns his gaze to two that are frequently absent from it: poverty and class."   —*Salon.com*

"The book doesn't lend itself to sound bites and should probably be at the top of any foodophile's reading pile. . . . [*The Gospel of Food*] the kind of work that, like *The Omnivore's Dilemma*, the food community will be talking about for a long time to come."   —*Chow.com*

"Barry Glassner has made it his business to set credulous consumers of mass media straight. . . . A master at the art of dissecting research, he points out that the obesity epidemic . . . came about through a complex mix of genetic predisposition, economic hardship, and antismoking campaigns."   —*New York Times Book Review*

"[Glassner's] is a plea for sanity, an exhortation to avoid the looniness of 'foodie' extremism. . . . Glassner's views about food are on the whole so bracing . . . it's hard not to root for him as he tilts against modern food dogma."   —*Wall Street Journal*

"In this fact-saturated sermon, Glassner argues that what we read in the health and food sections of newspapers . . . what we think we know about food . . . is wrong, biased, and based on limited or faulty research. . . . If Glassner preaches anything in *The Gospel of Food*, it is not to trust anybody's pronouncements about what to eat, be they from a scientist, nutritionist, or well-respected diet guru."   —*Washington Post*

"Glassner understands that any food can be made to seem good or bad depending on which features one chooses to emphasize, and that a dish's merits cannot be fully captured by a table listing vitamins, minerals, protein, fiber, fats, cholesterol, sodium, and sugar. Glassner is not oblivious to health concerns, but he points out when they are exaggerated or mistaken. . . . Glassner resists turning every meal choice into a moral statement or political act, and he is quick to question all forms of food snobbery, including his own. . . . In short, Glassner appreciates food in all its amazing variety and is not willing to deny what his palate tells him for the sake of fashion or ideology." —*Reason* magazine

"In his convincing *Gospel of Food*, Glassner looks at conflicting myths about food, such as the suggested health benefits of the Atkins diet, and the purported deadliness of eggs and hot dogs. Glassner decries those who preach 'the gospel of naught,' the idea that 'the worth of a meal lies principally in what it lacks.' He thinks America's obesity epidemic has been exaggerated, in part by a food industry eager to sell higher-priced 'natural' products, many of which have no more nutritional value than processed foods. The right path, he says, is to learn to take genuine pleasure from your meals. You'll be happier, which in and of itself will make you healthier." —Bloomberg.com

"For anyone who has longed for a good read that speaks sense to America's preoccupation with diet fads and irrational food fears, Barry Glassner is your man. Glassner made the rounds, talking to nutritionists, chefs, restaurant critics, physicians, and food chemists to divine the truth about why we're so afraid of food. . . . Glassner makes a highly readable argument for why everything you think you know about food is wrong." —*Denver Post*

The Gospel of
FOOD

The Gospel of

FOOD

*Why We Should Stop Worrying
and Enjoy What We Eat*

Barry Glassner

AN BOOK

HARPER  PERENNIAL

NEW YORK • LONDON • TORONTO • SYDNEY

HARPER ● PERENNIAL

A hardcover edition of this book was published in 2007 by Ecco, an imprint of HarperCollins Publishers.

P.S.™ is a trademark of HarperCollins Publishers.

HarperCollins books may be purchased for educational, business, or sales promotional use. For information please write: Special Markets Department, HarperCollins Publishers, 10 East 53rd Street, New York, NY 10022.

FIRST HARPER PERENNIAL EDITION PUBLISHED 2008.

Designed by rlf design

Library of Congress Cataloging-in-Publication Data is available upon request.

ISBN: 978-0-06-050122-8 (pbk.)

08 09 10 11 12   WBC/RRD   10 9 8 7 6 5 4 3 2 1

For

Sita and Jan Haldipur

Delaney, Megan, and Samantha Glassner

Ben and Leah Rafferty

Contents

Preface:
Eating Is Believing

"When you're born in this world, you're given a ticket to the freak show. When you're born in America you're given a front row seat," George Carlin, the comedian, observed in an interview with the *New York Times*.[1]

Having spent the better part of the last five years studying American foodways, I have to say, when it comes to how we eat, we're pretty freaky. I've talked with people who routinely eat foods they dislike because they wrongly imagine that by doing so, they will live longer or impress others. Sword swallowers at least get paid for their trouble.

One hot Sunday afternoon in Los Angeles, I found myself at a birthday party for a four-year-old with a bite of birthday cake spackled to the roof of my mouth. As I learned when the parents bragged of the "healthy recipe" they had used, the concoction contained none of the ingredients that make baked goods palatable. Though neither the parents nor their child nor any of the guests were vegans or celiac-disease sufferers, the cake had no eggs, butter, milk, or wheat. And needless to say, it had no sugar.

This couple, like many other twenty-first-century American parents, fear that if sugar passes their child's lips, she will turn hyperactive or diabetic, or both.

Every society has had its food preferences and prohibitions, usually dictated by religious teachings: Judaism and Islam prohibited pork, Catholicism decreed fish on Fridays. The difference today is that for huge numbers of people, *eating* is a religion. We worship at the temples of celebrity chefs. We raise our children to believe that certain foods are good and others are bad. We engage in elaborate rituals in preparing meals at home and describe ourselves as sinful if we order a creamy dessert when we eat out.

We even believe in miracles. In recent surveys, nine out of ten Americans said they believe that certain foods have benefits that go beyond basic nutrition. Exactly *which* foods and benefits people believe in varies greatly, however. Vegetarians believe their meatless regimen can prevent almost every serious malady from heart disease to world hunger. Followers of the late Dr. Atkins devour meat at every meal, persuaded that protein is a magical potion for weight loss and longevity.[2]

Other people base their faith less on the foods they eat than on where and with whom they eat them. In hopes of keeping their marriage together and their kids out of trouble, they encourage everyone in their household to come together around the dining room table. Or they never eat at home; they eat only at restaurants extolled by the gastronomic elite of which they aspire to be a part.

The good news about our food-obsessed age is the quality and variety of foods that have become available and the delight many Americans take in exploring new tastes. Yet even as many of us have embraced the pleasures of the table, many others, like the hosts of the birthday party, have fallen under the sway of killjoys who preach a *gospel of naught*—the view that the worth of a meal lies principally in what it lacks. The less sugar, salt, fat, calories, carbs, preservatives, additives, or other suspect stuff, the better the meal.

Without ignoring that consuming too much of anything is unhealthy, or that people with particular health problems need to steer clear of certain foods, we would all do well to maintain a healthy skepticism about the presumed sanctity and safety of one food or diet over another. Consider fruits, vegetables, and fish. There's no denying the oft-repeated admonition that to re-duce our odds of disease, we ought to consume more of them. But are they invariably safer than meat? From the many news reports about contaminated beef and poultry, you'd certainly think so. Beyond the headlines, though, are statistics from an investigative unit of Congress, the U.S. Government Account-ability Office, that tell a different story. Eight out of ten cases of food poisoning come not from meat, but from fruits, vegetables, seafood, and cheese.[3]

That study and others suggest we're more likely to become sick from what we eat today than we were fifty years ago in large measure because we're eating more raw fruits and vegetables. Our parents and grandparents cooked their veggies and skinned their fruits, thereby eliminating bacteria and viruses. They went overboard sometimes and fed us mush, but present-day com-mandments for how and what to cook come with costs and con-tradictions of their own.

Flash back to 1969, when *Gourmet* ran a cartoon that depicted two chefs conversing in a kitchen. "Henri," read the caption, "what went wrong? The broccoli tastes just like broccoli." At the time, great chefs were expected to transfigure ingredients into something greater than themselves. Today chefs are still ex-pected to perform miracles, but of precisely the opposite order.

Now the goal is "food that tastes like what it is," as Leslie Brenner, a food writer and editor, puts it. Chefs must prepare sophisticated dishes while retaining the distinctive flavors of the principal ingredients.[4]

Today, we demand that the fish we eat this evening swam with his mates last night and the vegetables that accompany him slept snugly in organic beds until early this morning. Believing we're extending our lives or the sophistication of our palates, we traipse eagerly, at inconvenient hours, to specialty shops and distant markets, places that our parents and grandparents, for equally lofty reasons, deliberately shunned. Recalling the drudgery that *their* parents endured to gather the component parts of the evening meal, they relished the chance to feed their families for an entire week with the bounty of one trip to a supermarket. Kroger's frozen or canned, preservative-protected foods seemed pretty miraculous to them. These foods also tasted wholesome back then because they embodied two of the prime values of the twentieth century: efficiency and technology.

As our values have changed, so have our views of food and the food industry. Nowadays, rather than put excessive faith in technological solutions and big food companies, we overvalue products labeled "natural" or "organic." In experiments where people are told they have been given those types of foods, they tend to rate them as tasting better, even when the researchers actually give them conventionally grown foods.

Psychologists call this "expectancy confirmation." The term refers to our tendency to find ways to fit experiences into the preconceptions we bring to them—something we do often with food. If a restaurant reviewer we admire likes a particular eatery, we are inclined to find merit in what it serves. If a nutrition newsletter from Harvard or Berkeley tells us a certain dietary supplement will improve our mood, we feel happier when taking it.[5]

Rather than let our palates be our guides, we let others tell us what to eat and how to think about what we eat. This book is about those others: the nutrition reformers, food critics, and other supposed experts who deify certain foods and eateries and demonize many others. The biggest beneficiaries of their exhortations, ironically enough, are the food companies. As I will show, every time a particular food gets singled out as the cause of corpulence or disease, the food industry, rather than losing money from declining sales, finds ways to make even more. The big losers are we eaters, who pay a premium for newfangled no-fat, low-carb, sugarless foods we don't really like.

It's time we learn to separate the wheat from the chaff. Only by recognizing the myths, half-truths, and guilt trips about what and where we eat can we begin to liberate ourselves for greater joy and realism at the table.

False Prophets

Culinary Correctness Gone Awry

T he word "enjoy" appears in the official dietary guide-lines issued by the governments of Britain, South Ko-rea, Thailand, and Australia. Norway comes right out and declares, "food and joy = health." The United States' dietary guidelines, faithful to our Puritan roots, say nothing about en-joyment.

It's high time we correct that omission. People get more out of a meal, not just emotionally, but physiologically, when the food is a pleasure to eat. In one of my favorite studies, Swedish and Thai women were fed a Thai dish that the Swedes found overly spicy. The Thai women, who liked the dish, absorbed more iron from the meal. When the researchers reversed the experiment and served hamburger, potatoes, and beans, the Swedes, who like this food, absorbed more iron. Most telling was a third variation of the experiment, in which both the Swedes and the Thais were given food that was high in nutrients but consisted of a sticky, savorless paste. In this case, neither group absorbed much iron.[1]

Similarly, studies of dieters find that those who regard plea-sure as unimportant in their food choices enjoy their meals less and are more likely to be dissatisfied with their bodies and ex-hibit symptoms of eating disorders.[2]

Then there's the much discussed French Paradox, the fact that the French eat a lot of what Americans believe will kill them, yet they die of heart attacks at about the same rate. The standard explanation, made famous in a *60 Minutes* segment in 1991, credits wine drinking, but there's surely more to the story. Serge Renaud, who first brought the paradox to light and runs a research institute at the National Institute of Health in Bor-deaux, suggests that another part of the answer lies in the types of fat the French consume. Goose and duck fat may do for the French what olive oil does for southern Europeans, Renaud hy-pothesized, namely, elevate their HDL ("good cholesterol"). Gascony has the lowest rates of heart disease in France, he noted, and people there eat a fair amount of foie gras.[3]

As my grandmother used to say, from his mouth to God's ear. Imagine if our risk of heart disease dropped with every bite of the sautéed Moulard duck foie gras with pickled white nec-tarines, onions, and arugula that my wife and I feasted on at the French Laundry on a recent birthday, or the poached foie gras with a marmalade of greengage plums that I will never forget from another dinner there.

More likely, any benefit to our health from Thomas Keller's wondrous cooking resulted from the happiness it brought us. Renaud's critics have appropriately pointed out that Gascons probably have other traits in common that protect them from heart attacks besides a fondness for foie gras.

One candidate is their attitude toward eating. Paul Rozin, a psychologist at the University of Pennsylvania, organized a study in which 1,281 people in France, Japan, Belgium, and the

United States were questioned about their attitudes toward food. Among the findings: the French view food as pleasure, while Americans worry about food. Asked what words they associate with chocolate cake, the French chose "celebration" and the Americans chose "guilt." Asked about heavy cream, the French selected "whipped"; Americans chose "unhealthy."[4]

We Americans see pleasurable and healthy eating as mutually exclusive. In a survey in 2000, *Newsweek* asked readers whether they consider long-term health when planning their diet. The four answers the magazine offered—"yes," "not as carefully as I probably should," "I pay more attention to my weight," and "no, I enjoy life and eat what I want"—show how dreary and dichotomized the American view of eating has become.

The results of the survey underscore the point. Fewer than one in four respondents selected "enjoy life."[5]

A Satisfaction Not Easy to Attain

We get that joyless view from nutrition writers and scientists who extol self-denial as a key to good health. "Spoil your appetite," Walter Willett, chair of the Department of Nutrition at Harvard, advises readers of his book on food and health. Have a snack of carrots or whole-grain wafers prior to mealtime, he recommends, so you don't eat as much at the table and risk gaining weight.

Willett's book overflows with pleasure-busting suggestions. He proposes, for example: "You may eat less if your entire meal is a chicken dish and vegetables than if you prepare several tempting dishes."

The very title of Willett's book announces that happiness is beside the point. *Eat, Drink, and Be Healthy,* he named it.[6]

Willett follows in a long line of nutritional scientists who have regarded the pursuit of pleasure at the table as either immaterial to good health or downright detrimental. Writing in a popular magazine in 1902, a physician looked forward to the day when "man has conquered his palate and no longer allows it to dictate the quantity and quality of the things he swallows." In *Literary Digest* in 1913, a chemist went further still. "It would be a hundred times better if foods were without odor or savor, for then we should eat exactly what we needed and would feel a good deal better," he declared.[7]

Present-day proponents of the doctrine of naught would banish some of nature's swellest edibles from our tables. The humble potato, for instance, is "part of the perilous pathway to heart disease and diabetes," according to Willett. A baked potato may look innocent enough, but it turns to glucose, he cautions, which produces dangerous surges in blood sugar and insulin. Instead of classifying potatoes as vegetables and encouraging people to eat them, as it currently does, the U.S. Department of Agriculture should call them carbohydrates and have us consume them only occasionally, Willett maintains.[8]

Never mind that the potato was the principal source of sustenance in Ireland in the late eighteenth century and the early nineteenth century, providing most of the calories, protein, and vitamins that kept the peasant population alive. "Hard as is the fate of the labouring man, I think he is greatly indebted to the potato for his flow of spirits and health of body," wrote Asenath Nicholson, an American diet reformer who spent four years in Ireland in the 1840s and published her observations in a book titled *Ireland's Welcome to the Stranger.*[9]

Historians credit the potato with having made possible the population upsurge in central and northern Europe in the 1700s and 1800s, and as recently as the late 1990s, in a bestseller, *Pota-*

toes Not Prozac, this versatile tuber was being heralded as a cure for depression. But in 2002, *Time* magazine ended a feature article about foods that purportedly prevent disease with a sidebar titled "And Now the Bad News: Potatoes." The piece quoted one of Willett's colleagues in the Department of Nutrition at Harvard explaining that potatoes push down good cholesterol and drive up triglycerides.[10]

Pity the poor soul who took Willett and company's advice and swore off spuds. "Baked slowly, with its skin rubbed first in a buttery hand, or boiled in its jacket and then 'shook,' it is delicious," M. F. K. Fisher, the eminent food essayist, rightly wrote. "Alone, or with a fat jug of rich cool milk or a chunk of fresh Gruyere, it fills the stomach and the soul with a satisfaction not too easy to attain."[11]

On Joel Robuchon's list of the eight primary ingredients in his cuisine, the potato appears alongside caviar, scallops, truffles, crepes, sweetbreads, chestnuts, and almonds. Of the many extraordinary dishes at Jamin, Robuchon's legendary three-star Paris restaurant that closed in 1996, the potato puree is the most famous. Made from the finest butter (a great deal of it, eight ounces for every pound of potato) and *la ratte,* an heirloom potato with a hazelnut flavor, Robuchon's mashed potatoes changed lives. In conversations with food enthusiasts in the nearly twenty years since I tasted that dish at Jamin, I have discovered that I am far from the only person who credits that potato puree for a lifelong interest in great cooking.

Forgo potatoes? Better to follow the lead of some of the greatest chefs of our age and find ever more inventive uses for them. "I think potatoes are a magic ingredient," Michel Richard told me during an interview one morning in the kitchen at Citronelle, his celebrated restaurant in Washington, D.C., and afterward, a friend and I had lunch at the restaurant and Richard proved his point. Each of the five courses he sent us, save the dessert, included potatoes. Buttery whipped potatoes lay beneath the foie gras. Crispy potatoes encircled the tuna tartare. Cumin-flavored potato chips accompanied the squab.

And there was the potato course itself, a glorious risotto made of Yukon Golds cut into eighth-inch dice and cooked to the texture of rice in a creamy, garlicky broth. Somewhere around his third bite, I saw on my friend's face the look of revelation I must have had at Jamin. He'd had no idea, he told me, that potatoes could be as good as sex.

Walking on Eggs

The truth be known, they're good for you too. "Potatoes are a fat-free, sodium-free and cholesterol-free food. They are high in vitamin C and potassium and provide a good source of vitamin B6 and dietary fiber," notes a brochure published by Oldways Preservation Trust, a food advocacy group based in Boston.

A close look at the brochure reveals one of the secret failings of the church of naught. Though they like to pretend otherwise, the priests disagree among themselves. On the flip side of this pro-potato message is a picture of none other than Walter Willett. He is one of four "top nutrition experts" who provided blurbs in support of the Mediterranean Diet Pyramid that Oldways has been hyping and that calls for less meat, more plant and vegetable oil, and plenty of potatoes. ("Get into a Mediterranean frame of mind by creating quick and easy potato-based meals," the brochure urges.)

Willett's blurb does not say anything about potatoes per se, but the fact that there could be a head shot of an anti-potato man on the reverse side of a pitch for potatoes speaks volumes about the folly of most dietary mandates. Oldways and like-minded organizations go to great lengths to devise clear and consistent lists of what the public should and should not eat. They convene special conferences and call together panels of experts from around the world for that express purpose. Scrutinize their proclamations, however, or the process that produced them, and you find a gumbo of conflicts, contradictions, politics, and personalities.

"Like most contemporary issues, the question of what people should eat in order to maintain good health cannot be neatly split into matters of 'fact' and matters of 'values,' " Stephen Hilgartner, a professor in the Department of Science and Technology Studies at Cornell, concluded after a detailed study of three influential reports on diet and health from the National Academy of Sciences. "The question, 'What are the facts?,' is entangled in questions about the criteria for determining facts, which in turn are connected to questions about who can be believed, which institutions are credible, what scientific methods are reliable, and how much evidence is needed to justify altering the status quo."[12]

Committees of experts convened by the government sometimes are unable to resolve those questions or reach agreement about particular dietary advice. When they *do* achieve consensus, a great deal of what Hilgartner politely refers to as "information control" and "persuasive rhetoric" is involved. The committee's chair sees to it that differences of opinion among committee members are kept within bounds and the views of qualified scientists who disagree with the official statement don't muddy the waters. Skeptical scientists are excluded from the committee in the first place, or their concerns are downplayed in a carefully worded consensus report.[13]

Advocacy groups have to be political too in deciding which foods to damn in their public documents. Had Oldways moved potatoes to the condemned end of the food pyramid, the group's leaders would have avoided the embarrassment of contradicting Willett, but in so doing, they would have broken ranks with another of their well-known supporters. Jane Brody, the *New York Times* health columnist and a frequent speaker at Oldways meetings, had been championing potatoes as nutritious and nonfattening for the past quarter century.

In 1981, when *Jane Brody's Nutrition Book* came out, she told an interviewer that her husband has been "a confirmed meat-and-potatoes man with the emphasis on meat" before he began typing the manuscript for her book. "But in the process of typing," she said, "he became a convert to my nutritional philosophy, so now he's a meat-and-potatoes man with the emphasis on the potatoes."[14]

No enemy of potatoes is a friend of Jane Brody's. During an interview, when I asked her about Walter Willett's recommendations about them, she could barely contain her ire. "His research may be fine, but his opinions really leave me cold," she replied. The fact that a potato can raise blood sugar is "absolutely meaningless," she says, because people don't eat potatoes by themselves. "A Mars bar is likely to be eaten that way, but how many baked potatoes have you eaten without anything else? I'll bet none. When you eat the potato in the context of a meal, it doesn't do the same thing because it's mixed with many other nutrients, which neutralizes the glucose-raising effect of the potato."[15]

When she finishes her defense of potatoes, I ask about the item that had accompanied my hash browns at breakfast. Eggs, I remind her, were on the condemned-foods list not long ago. To those of us who like to eat, Brody's more than one hundred articles over the years cautioning against a lengthy list of foods, from meats (except on special occasions and then only lean and skinless cuts smaller than four ounces) to Girl Scout cookies ("they should be banned from the face of the earth," she declared at an Oldways conference) have been excessive, to say the least. But she makes light of the fact that she and fellow preachers of the gospel of naught have been known to damn a food one year and absolve it the next.[16]

"I use that Humpty Dumpty analogy," Brody replies. "The egg had a very bad rep, but we are putting Humpty Dumpty back together again. I've resurrected the egg and so has the American Heart Association. The Heart Association now says the vast majority of people can eat one egg a day without any problems. But if you happen to be a person whose body is sensitive to dietary cholesterol, and that may be as much as 10 percent of the population, then you have to be careful about those high-cholesterol foods."

When her twin boys were young, she frequently fed them a dish she called "Eggs Jane"—an English muffin, a slice of turkey breast, poached egg white, and a slice of cheese heated in a toaster oven. "It was my version of Eggs Benedict, but without the bad things," she says, adding that nowadays she eats yolks.

Back in the 1970s and 1980s, Brody had been among the chorus of nutritionists and health writers who warned the American public that yolks are packed with cholesterol. The more cholesterol a person ingests, the theory went, the higher his blood cholesterol and the greater his odds of a heart attack, so eggs were seen as potentially lethal. And as a result of the bad publicity, egg consumption in the U.S. plummeted.[17]

Yet Brody says she does not regret having discouraged people from eating eggs. In her view, she was merely reporting the state of knowledge at the time. "We know something that we didn't know then," she says. "We now know how important HDLs are. When we only looked at LDLs, the bad cholesterol, it looked terrible to eat a lot of eggs. But if you look at the ratio of total cholesterol to HDL, the good cholesterol that cleanses your arteries like Drano, then for a lot of people who thought they couldn't eat eggs, eggs are okay."

The Wit to Eat Whole Yogurt

I have learned to expect this type of rejoinder whenever I ask defenders of the doctrine of naught about the considerable quantity of crow they have had to eat for vilifying foods that deserved better. *We had no way to know,* they insist, when in actuality, there was reason to withhold their negative judgment. At the time they were castigating the egg, for example, not a single study had shown that eating eggs produced higher rates of heart disease.[18]

Nor has research since that time. On the contrary, nutritionists have long known that eggs have much to commend them as staples in the human diet. They supply protein, B vitamins, and other nutrients at low cost, and although they're rich in flavor, they go well with a range of other ingredients in recipes. What's more, they're amenable to just about every major method of cooking, from boiling to baking to frying, and to more textures and shapes, from foamy to fluffy to firm, than any other food.[19]

Eggs were one of many victims of an unfortunate campaign against dietary cholesterol that began in the 1950s and still continues by virtue of the government-mandated nutrition label on every food product. Since 1994, food manufacturers have been required to include in their packaging a table of "Nutrition Facts" that directs consumers to restrict total daily cholesterol to little more than the amount in one large egg. Even though Jane Brody has come to acknowledge that most people can con-

sume more than that amount of cholesterol with little or no adverse effect on their blood cholesterol levels or their hearts,
cholesterol remains one of five demonized nutrients on food labels, along with saturated fat, sugar, sodium, and trans-fatty
acids.[20]

Rather than call all of that to Jane Brody's attention during
our interview, I raised a more general question. I asked if she
receives much criticism of her views about food.

She said no, but then told a story that helped me understand
the appeal of her philosophy of food for large numbers of people. "Somebody did just ask me," she recounts, "how I can recommend fish when fish has all this stuff in it that's not good for
you, mercury and I don't know what else. I said, 'Well, we have
to eat every day, and we have to make choices, and everything
has a downside.' If you eat four hundred carrots in one day,
you'll probably die because four hundred carrots are poisonous.
That doesn't mean you can't eat two or three carrots."

Some of us see eating as something we *get* to do, a privilege
and source of joy. Others view eating as something they *have* to
do. For those who take the latter view, Brody's columns are the
journalistic equivalent of Powdermilk Biscuits. Like Garrison
Keillor's fictitious product, they "give shy persons the strength
to get up and do what they have to do." Her columns embolden
them to eat what they ought to eat and sidestep the rest.

"You're either a fat slob who eats junk all day, or you're a
perky person who weighs out bits of skinless chicken and drinks
low-fat milk," Emily Green, a feature writer at the *Los Angeles
Times*, paraphrases Brody's view of the public. "What's wrong
with cooking and eating and living in a fulfilling way that is
actually conducive to healthfulness? What's wrong with having a good meal and working in the garden in the afternoon?"
asks Green, who did precisely those things on the afternoon I

interviewed her at her home. She prepared a resplendent lunch for us, and after our meeting, returned to work in her backyard garden, an urban Eden of green and flowering plants that attracts many-colored butterflies and birds.

Relatively little of our lunch complied with the dietary advice Green lambastes. The spinach, steamed and dressed in extra-virgin olive oil from Nice, conformed, but not the omelet stuffed with cheese, and certainly not the whole-milk yogurt from Straus Family Creamery, a small producer located near San Francisco. Lusciously thick, it is infinitely more satisfying than the nonfat yogurts that monopolize the dairy cases of American supermarkets.

Green has written eloquently against nonfat dairy products, low-fat sweets, and the like—products she christens "nonunde-lows" because their names begin with non-, un-, de-, or low. "In a superb sleight of hand, we have been led to believe that the leaching of those pesky 'nutritive' elements from our food and drink is somehow good for us," Green has written. She argues, to the contrary, that the health benefits of nonundelows remain unproved, and their proliferation has contributed to a rise in obesity. "Yes, of course there are other factors in the fattening of America: We drive more and walk less, and so on. But my own guess," she says, "is that we can't stop eating because nonunde-lows leave us hungry."[21]

Partly to prove her point, Green put herself on a weight-loss diet that included no nonundelows. "To my knowledge, not a single low-fat food passed my lips," she says of the diet and exercise plan on which she lost fifty-two pounds in fifty-two weeks without denying herself what she calls "nonnegotiable plea-sures" like meals at great restaurants.[22]

Green did cut back on some things. She limited her candy intake to one Valrhona dark chocolate a month, wine to one glass with dinner, and bread to a few times a week (garlic bread at a beloved lunch place and the bread basket at Campanile Restaurant). The only foods Green gave up entirely were pasta, cookies, doughnuts, and nonundelows.

She assails others journalists who do not "trust us with real food, with the wit to eat whole yogurt, real milk or cream when we want something filling, and an apple when we want a light snack." The very mention of low-fat or nonfat milk enrages her. "It's thin and nasty and has all of its goodness stripped out of it," she says when I broach the subject. "If there's too much fat in something, have less of it. I don't go with the idea, 'Oh, Picasso's canvases are too big, let's just cut them in half.' "

That sort of reasoning, along with the "One Percent or Less" low-fat milk campaign initiated by the Center for Science in the Public Interest, an advocacy group, and promoted by assorted nutrition writers and government agencies, has resulted in abandonment of whole milk in the U.S. Consumption has decreased by two-thirds over the past three decades, while sales of low-fat and nonfat milk have more than tripled.[23]

Where Hot Dogs Trump Spinach

Many Americans, including some serious gastronomes and accomplished home chefs, have not tasted whole milk in years. A few weeks after my visit with Green, at a special dinner of Turkish food presented at a Los Angeles restaurant, the first item brought to our table was a large bowl of what looked and tasted like a rich, creamy, exotic pudding. My wife and the couple who joined us—all practiced cooks and savvy diners—scoffed when I identified it as yogurt. I would have been in the dark myself had it not been for my visit with Emily Green, a fact I chose not to reveal. I opted for smug silence when Evan Kleiman, the feast's chef, came to our table and responded to my wife's query about the mystery dish. Americans are so unaccustomed to full-fat yogurt, Kleiman said, that it tastes spectacular to us.

What's more, dairy fat contains conjugated linoleic acid (CLA), which studies suggest inhibits cancers of the colon, breast, and stomach and decreases the risk of heart attacks. But devotees of the doctrine of naught get hardly any CLA in their diets unless they lapse and partake of verboten foods that contain it. Which raises a question: If future studies confirm the benefits of CLA, ought nutrition writers to urge Americans to swear off skim milk and eat ice cream?[24]

Personally, I would love an excuse to consume Ben & Jerry's Cherry Garcia, which does wonders for my soul. Sadly, though, the same cannot be said for my body. Half a cup of Cherry Garcia delivers 260 calories, as much sugar as a Mr. Goodbar, and a hell of a lot of saturated fat. You don't want to eat too much, CLA or no CLA.

No one would seriously advocate ice cream as a health food, though in fact that advice is no less fallacious than its opposite, a faulty logic that assumes if a steady diet of something is harmful, going without it must be healthful. That wrongheaded reasoning is rampant. For one of his studies, Paul Rozin presented the following scenario to a diverse sample of Americans: "Assume you are alone on a desert island for one year and you can have water and one other food. Pick the food that you think would be best for your health." Seven choices were offered: corn, alfalfa sprouts, hot dogs, spinach, peaches, bananas, and milk chocolate.

Fewer than one in ten people chose hot dogs or milk chocolate, the two foods on the list that come closest to providing a complete diet because of the fats and other nutrients they contain.

In response to another set of questions, half of Rozin's respondents said that even very small amounts of salt, cholesterol, and fat are unhealthy. More than one in four believed that a diet totally free of those substances is healthiest, when in reality, of course, they are crucial for human health. Without them, we could not survive.[25]

Most nutrition writers are not likely to correct those misconceptions. Their goal is not to elucidate the virtues of hot dogs, fats, and seasonings, but rather, as Emily Green put it, "to keep nasty food out of people's mouths." Nor is there much incentive for other journalists to challenge the conventional wisdom. Those who do typically find themselves accused of being an enemy of public health.

Green and other doubters routinely receive caustic criticism from advocacy organizations such as the Center for Science in the Public Interest and from readers who accuse them variously of ignorance, naïveté, or duplicity. After Gary Taubes, a veteran science writer, argued in a *New York Times Magazine* piece in 2002 that foods such as steak and cheese, "considered more or less deadly under the low-fat dogma, turn out to be comparatively benign," and that "cutting back on the saturated fats in my diet to the levels recommended by the American Heart Association would not add more than a few months to my life," he was decried by spokespeople for advocacy and governmental organizations, and more vehemently still, by nutrition writers. Jane Brody devoted an entire column to repudiating him, and in a phone interview I had with her the day after Taubes's article came out, she dismissed his claims as "total conjecture" and "irresponsible."[26]

"Laughable" was the word that Sally Squires, the nutrition columnist at the *Washington Post*, chose to describe Taubes's argument, even though an earlier version had been published in *Science* magazine and won a National Association of Science

Writers award. "Get real," Squires demanded in one of three disapproving pieces she published soon after Taubes's article appeared. Recalling Woody Allen's 1973 movie *Sleeper,* in which a man wakes up two hundred years hence and informs doctors that steaks and cream pies were once believed to be unhealthy, Squires counseled readers: "We laughed then, and we should be laughing now."[27]

Those are pretty harsh words from one Columbia Journalism School graduate about another, especially considering that Taubes does not actually deny the central tenet of the doctrine of naught. Neither in his articles nor in his book does Taubes disagree that whole classes of delectable and nutrient-rich foods should be largely eliminated from the American diet. Instead he demonizes a different group of foods, "those refined carbohydrates at the base of the famous Food Guide Pyramid—the pasta, rice and bread—that we are told should be the staple of our healthy low-fat diet." In Taubes's scheme, eggs at breakfast are fine, but hold the toast. Steak dinners can't be beat, assuming you skip the potatoes.

(Taubes eats what he preaches, by the way. He and I have had several lunches together at restaurants that serve warm, fragrant breads before the meal, and with my main course, pan-fried red-skinned potatoes or homemade pasta. Yet Taubes feasts away on his meat, unmoved by the pleasures on my plate.)

A Lone Voice of the New York Times

In the health or science section of a major American newspaper, you are about as likely to find a vocal skeptic of standard dietary advice as you are an anticapitalist in the business section. Critics like the *Los Angeles Times's* Emily Green write for the "food" and "home" sections, and freelancers such as Gary Taubes publish occasional pieces in the *New York Times Magazine*. To my knowledge, only one full-time science and health reporter—Gina Kolata at the *New York Times*—has dared to dispute the doctrine of naught.

Kolata's byline appears on most of the *Times* pieces over the past decade that have raised doubts about the wisdom of vilifying particular foods:

"Benefit of Standard Low-Fat Diet Is Doubted"

"Scientists Cautious on Report of Cancer from Starchy Foods"

"In Public Health, Definitive Data and Results Can Be Elusive"

"Amid Inconclusive Health Studies, Some Experts Advise Less Advice"

"The Body Heretic: It Scorns Our Efforts"

In those and other articles over the past decade, Kolata has questioned whether low-fat and low-cholesterol diets reduce the incidence of heart disease and cancer; whether eating sugar causes obesity; and whether consumption of acrylamide, a much-maligned chemical in French fries and other starchy foods, causes cancer.[28]

The moral of many of Kolata's stories was summed up in an article she published in 1999. "Sometimes well-intentioned advice is later revealed to be based on hopes rather than facts," wrote Kolata, who contends that studies of the relationship between diet and health have serious limitations. Most are observational: they survey people's eating habits to see if those who develop heart disease or cancer have different diets than those who stay well. "Such studies have a fundamental drawback," Kolata points out. "People who eat in a particular way are very different than those who don't eat in a particular way."[29]

Those who subscribe to the doctrine of naught may have different personality traits, genetic predispositions, job or family pressures, or leisure-time activities than people who eat Big Macs. And any of these nondietary differences may account for differences in rates of heart disease or cancer.

Even the largest and most highly cited observational studies have had both hits and misses. Kolata notes that the Nurses Health Study, directed by Walter Willett, correctly identified smoking as a cause of cancer and heart disease but wrongly concluded that hormone-replacement therapy protects against heart disease, Alzheimer's, and osteoporosis, and that vitamin E protects against heart disease. Prominent university researchers have raised similar concerns about Willett's study.[30]

In principle, the problems of observational studies can be avoided by either of two alternative approaches—international comparisons or randomized trials. But similar difficulties arise in those as well. Are the lower rates of heart disease in China and Japan attributable to diets low in saturated fat or in refined carbohydrates, as proponents of the gospel of naught contend, or to some of the many other cultural and culinary differences between those societies and ours? And how reliable are estimates of heart disease and eating habits in China? Obtaining dependable information is difficult enough in our own nation of 288 million inhabitants, never mind in a country with more than four times as many people, many of whom live in rural areas with lower rates of literacy.[31]

Optimally, the effects of diet would be assessed the same way drugs are tested, through experiments in which some people are given the substance in question while others take sugar pills, and neither group knows which it has received. Obviously, that sort of study is close to impossible when it comes to diet.

You can't give one person a T-bone and another tofu and have them believe they are eating the same thing. The closest that nutrition researchers come to randomized trials are experiments in which they assign people to eat particular foods rather than give them free choice. That scheme reduces the likelihood that other commonalities among people are responsible for differences in their rates of disease. But as Gina Kolata and others have pointed out, such experiments are difficult to pull off successfully because people have a hard time sticking to mandated diets.[32]

Eating Isn't Smoking

The consequences of errors are potentially greater in studies of diet than in other sorts of research on how our behaviors affect our health. If, because of faulty information, a study concludes that smoking increases the risk of lung cancer by only 2,000 percent rather than 3,000 percent (the amount other studies have shown), the moral of the story remains the same. Either way, there can be little doubt that smokers run a significantly greater risk than nonsmokers. By contrast, studies of the effects of foods on heart disease and cancer often show an increased risk of only 20 or 30 percent. A few errors in measurement and the danger disappears almost entirely.

In many of the studies, the number of people in the group that ate the purportedly unhealthy food and got sick is shockingly small. A report from the Nurses Health Study that appeared in the *New England Journal of Medicine* and led to a number of frightening news reports is a case in point. Women who eat beef, pork, or lamb every day have two and a half times greater risk of developing colon cancer over a six-year period compared with women who eat red meat less than once a month, Willett and his colleagues reported.

With findings like that, it is easy to appreciate why Willett responded so concisely to a question from a reporter. Asked how much meat people should eat, he gave a one-word reply: "Zero." But when I examined the journal article, I discovered that the "two and a half times" figure was based on a small number of women. A total of 150 of the 89,000 women in the study developed colon cancer over a six-year period. Of those who said they ate red meat less than once a month, 14 developed colon cancer, compared with 16 who said they ate red meat every day. If just a few women inadvertently misinformed the researchers about how much meat they ate or their health status, the front-page headlines about the risk of meat eating may well have been off base.[33]

A much-publicized discovery from a study about chili peppers almost certainly was erroneous. "If our findings are right, the risk of getting stomach cancer from consuming large amounts of chili peppers would be almost on the order of getting lung cancer from smoking," Robert Dubrow, a Yale epidemiologist, proclaimed after he and some collaborators from Mexico's National Institute of Health concluded that people who eat the most peppers are seventeen times more likely to get stomach cancer than those who eat none.[34]

When I looked at the study, I learned that of the 972 residents of Mexico City who were interviewed, only a small number had stomach cancer and said they ate a lot of peppers. The study's dramatic finding evaporates if just a few of these folks overestimated their chili consumption, as well they might. People tend to search for causes of their illnesses, and since peppers cause gastric distress, they may come to mind as candidates.

Reports from the self-described abstainers in the study strike me as suspect as well. Some of them may have consumed chili peppers without knowing or remembering they had, especially in a country where chili peppers are ubiquitous.

Getting accurate information on other people's diets is famously difficult because, as the sociologist Georg Simmel wrote a century ago, "what I think, I can let others know, what I see, I can let them see, what I say, hundreds can hear—but what the individual eats, no one else can eat under any circumstances."[35]

People often misreport what they eat not because they consciously want to deceive but because they are not paying careful attention or they think of themselves as eating another sort of diet than they really do. "The level of measurement error in food-frequency questionnaires is just so big, the results are very hard to interpret," John Powles, an epidemiologist at Cambridge University, told me.[36]

In a study like the one in Mexico City, the true findings may be impossible to interpret reliably. Soon after that study made headlines throughout North America, other researchers reported that chili peppers had the opposite effect. Hot peppers may actually reduce the risk of cancer by supplying antioxidants and neutralizing some carcinogenic substances found in other foods, laboratory studies found.

Epidemiologists chimed in as well. Stomach cancer rates do not tend to be higher, they said, in places where hot peppers are a regular part of the cuisine.

Indeed, just a few years later, one of the researchers on the Mexico City study reanalyzed the interview data and abandoned the peppers hypothesis. This time—with no better evidence than before—he fingered salty snacks, meat, and dairy products as the culprits.[37]

The Danger in Crying Wolf

To hear advocates of the doctrine of naught tell it, the scientific evidence is clear and decisive, and no scientist in his right mind seriously doubts the major tenets of the reigning view. In reality, however, as historian Felipe Fernandez-Armesto of Oxford University has written, "one of the few verifiable laws about dietetics is that the experts always disagree."[38]

In the course of my research, I discovered many well-versed scientists who challenge the conventional wisdom. Some, such as James Le Fanu of England and Uffe Ravnskov of Sweden, are physicians without academic positions, who take it upon themselves to study the scientific research and publish books with titles like *The Rise and Fall of Modern Medicine* and *The Cholesterol Myths*. Others are prominent researchers like Powles at Cambridge, whose findings turn the doctrine of naught on its head. He has identified groups of Greeks, Italians, and Japanese whose death rates from heart disease dropped as their meat consumption and blood cholesterol levels increased.[39]

Closer to home, I came upon articles by Marcia Angell, former editor of the *New England Journal of Medicine* and a senior lecturer in the Department of Social Medicine at Harvard Medical School. "Although we would all like to believe that changes in diet or lifestyle can greatly improve our health," Angell wrote in an essay in 1994, "the likelihood is that, with a few exceptions such as smoking cessation, many if not most such changes will produce only small effects. And the effects may not be consistent. A diet that is harmful to one person may be consumed with impunity by another."[40]

In an interview, Angell told me that the incessant warnings about foods may even do harm. "There is an analogy," she said, "to the story of the boy who cried wolf. If you're always ascribing things to diet and lifestyle, then when you do hear about something that's based on good evidence and really does have an effect, you've gotten cynical about it. You have just heard too much, often contradictory stuff, to take real threats seriously. A good example is cigarettes. They are a real threat, and yet, a lot of people look at smoking cigarettes as no worse than eating hot dogs."

I asked Angell why seemingly well-meaning epidemiologists and nutrition researchers would make the dangers of foods seem greater than they are. "They want grants and publicity," Angell replied. "Medical research is no longer done in an ivory tower. The National Institutes of Health and various companies that fund the research read the newspaper too. Publicity is very good for researchers."

Besides, she let me know, scientists who study the connection between food and disease may be believers themselves. "Researchers, even though they're supposed to be totally impartial, often carry with them sets of biases, and if they want to show something, they often work very hard to show that. They conclude what they want to conclude when there are other possible conclusions that would flow equally from their data."

Where, I asked Angell, does that leave the public? Are there foods that people really should shun?

"Within limits they should eat the way they want to eat," Angell replies. "What are the limits? I think they should eat in moderation, and I think they should eat as varied a diet as possible because that's good insurance. You don't put all of your eggs in one basket, or in this case, your health in one egg. You try to cover the waterfront because you're operating from a position of extraordinary ignorance, so your best bet is to eat a varied diet."

Eat what you want. I heard that advice not only from Marcia Angell, but from my personal physician as well. In fact, he had me conduct a little experiment that further strengthened his point.

His advice to eat what I want came initially during an annual physical exam, after he informed me that my cholesterol numbers had gone up compared with the previous year. I asked if I should change my diet, and in response, he took out a pad and scribbled a prescription for a cholesterol-lowering drug while instructing me—in a wearied voice, as if to a medical student who had asked a dumb question—that changing my diet probably will not help my heart or any other organ and I should eat what I want to eat.

Mild humiliation is a price I pay for having my medical examinations performed by Ricardo Hahn, of the Department of Family Medicine at the University of Southern California, where I work. All things being equal, I would not voluntarily put myself in the position of being corrected by a fellow professor while I sit naked except for my shorts and socks. But when it comes to medical care, things are never equal, and I prefer Hahn over docs who take at face value what they read in the health section of the local newspaper.

"What we think we know about nutrition is not supported by real scientific inquiry," he told me on a subsequent occasion. Little biological evidence exists, he said, to support the claims of those who caution against particular foods. If some people are healthier for having eschewed those foods, the reason may well be psychological. "Because of the placebo effect, people feel better when they adopt certain dietary habits," Hahn contended.

When I reminded him that his views are at odds with what one hears from physicians at the American Heart Association and diet advisory panels of the U.S. government, he recommended I do a small study myself to see how much those docs are really willing to attribute to diet.[41]

"You have to ask the right question," said Hahn. Ask general questions about whether diet matters, he advised, and you'll get platitudes in return. Force them instead to give a number—ask the percentage that diet contributes to particular diseases—and they'll sing a different tune.

Hahn encouraged me to test his hypothesis on the then-president of the American Heart Association, David Faxon, a cardiologist at the University of Chicago with whom Hahn had worked in the past. I should call Faxon and insist he be precise, Hahn instructed.

And sure enough, when I reached Faxon and asked him the percentage that diet contributes to heart disease, the question seemed to stop him in his tracks. "Wow. That's a very difficult question to answer, frankly," he said. "I guess part of the reason it's hard to answer is, you don't have as much information about the importance of diet on cardiovascular risk. We have a lot more information on some of the things that are affected by diet and the effects of drugs on those things, for instance, cholesterol."

Faxon went on to say that as a clinician he believes that diet matters more for some people than for others, and he made general statements endorsing the Mediterranean Diet Pyramid and the care with which AHA committees come up with their dietary guidelines. Then he acknowledged the relative lack of knowledge about the effect of diet on heart disease. "We have limited information in a number of studies on dietary modifications to cardiovascular risks," he said. "But when you compare it with the wealth of information that we have on other things, the data is really small."

Faxon in turn urged me to talk with Ronald Krauss, the principal author of the AHA's dietary guidelines. An M.D. and senior staff scientist in the Life Sciences Division at the Lawrence

National Laboratory in Berkeley, California, Krauss has served also on the Food and Nutrition Board of the National Academy of Sciences.

I reached Krauss by phone and asked him to estimate the percentage that diet contributes to disease, and he answered my question with a question. "Are you talking about all of diet and all of disease?"

I suggested we focus on his specialty, heart disease.

"I suppose to say the word 'important' is not enough," he tried, and I respectfully asked if he could be more specific.

"I don't even know where to start in trying to answer that question," Krauss said, audibly annoyed. After a little more prodding, he said that, on average, heart disease is attributable half to genetic factors and half to lifestyle factors such as diet, exercise, smoking, and body weight.

From having looked at some of Krauss's published papers, I know that much of his own research explores how genetic factors influence people's responses to diet. He and other researchers have documented that individuals react very differently to low-fat or reduced-salt diets, for example, depending upon their genetic predispositions. One person's LDL or blood pressure plunges, while someone else's remains nearly unchanged.[42]

Someday doctors might be able to estimate how much difference a particular change in diet would make for a particular person, on the basis of one's genetic profile and other information. But for now, Krauss told me, if I wanted an answer to the question I had posed to him, the best he could advise was for me to speak with epidemiologists, those who study patterns of disease in populations.

Not that Krauss himself has great faith in epidemiological studies, mind you. "The information on diet is rudimentary at best," he volunteered, and he characterized epidemiological data as "extremely erratic." Nonetheless, the largest such study, the Nurses Health Study, has collected "pretty incredible information," he said, and those researchers have come up with sophisticated estimates of the effects of diet on health.

I should talk with Walter Willett, Krauss suggested—and I heard myself thinking, *Of course*. If there is one person who will answer my question decisively, Willett is the guy. In fact, in his book, *Eat, Drink, and Be Healthy*, he had already come close to providing the number. "A healthy diet teamed up with regular exercise and no smoking can eliminate 80 percent of heart disease and 70 percent of some cancers," Willett proclaimed on the first page of the first chapter.

My Brilliant Diet

I reached Willett by phone, read that statement back to him, and asked him to break down the figure for the influence of diet alone.

You could have pushed me over with a French fry as I listened to his response. Even Walter Willett would not come up with an estimate. "Well, we haven't done exactly that. Smoking is the single most important factor," was his reply.

The best Willett could do was refer me to a paper he and his colleagues had published in which they removed smokers from their analysis and estimated the impact of diet and other factors for the nurses that remained. That paper includes a statistical table showing a 28 percent decreased risk of heart attacks among nonsmokers who said they did the following three things: consumed fish, fiber, and folate; exercised at least thirty minutes a day; and largely avoided saturated and trans fats and glucose-spiking carbs.[43]

Suddenly, I felt incredibly well protected against heart disease.

I'm a nonsmoker who has two of the three "low-risk" behaviors. Although I regularly partake of forbidden fats and carbs, my day begins with a brisk walk up steep hills, followed by a breakfast that includes plenty of fiber, as well as fruits that contain folate. And hardly ever do I go for more than a couple of days without a gorgeously baked or broiled fish, or sushi at my favorite place in Little Tokyo.[44]

Willett's paper suggests that were I to adhere to the doctrine of naught, I might reduce my risk of a heart attack by about 9 percent—an underwhelming number, especially when it is translated into actual heart attacks. According to the information presented in Willett's paper, there would have been two fewer heart attacks per thousand people over a fourteen-year period if the nonsmokers in the study who exercised and ate the recommended foods had given up the disapproved fats and carbs.

Actually, that number may be even tinier. In calculating it, I treated all three of the "low-risk factors" as equally important. More likely, the two positive factors have a greater impact than the negative one. After all, numerous studies have demonstrated large health benefits from exercise, and another professor of epidemiology at Harvard has shown that what we eat matters more than what we avoid. "It appears more important to increase the number of healthy foods regularly consumed than to reduce the number of less healthy foods regularly consumed," Karin Michels stated in a paper published in the *International Journal of Epidemiology*.[45]

Dangers Too Big to See

In a memorable old joke, a passerby comes upon a man who is searching for something beneath a streetlamp on a dark night.

The passerby asks the man what he's doing. "I lost my wallet a couple of blocks away," he replies. Perplexed, the onlooker asks why he doesn't look for it closer to where he lost it.

"The light's better here," the man explains.

Promoters of the doctrine of naught make certain that their lists of disapproved foods stay in the spotlight. That does not make them, however, the best places to look for the causes of chronic diseases. Having reviewed well over a thousand studies on the subject and spoken with a range of experts, I can suggest a list of likelier places.

Genes and the environment are the most obvious examples, and there are others. Viral and bacterial infections, job stress, living in distressed neighborhoods, early deficits such as malnutrition, low birth weight, or lack of parental support, and chronic sleep loss during adolescence and adulthood—none of these gets as much attention, but each has been shown to contribute to the development of heart disease, cancer, and other serious health conditions.[46]

According to Ichiro Kawachi, a professor in the Department of Health and Social Behavior at Harvard, the number one place to look for the causes of chronic diseases is the larger society. "The big social things like inequality, disparities in wealth and income, living conditions, and social cohesion explain 100 percent of the difference in cardiovascular disease across society," Kawachi told me. In his view, factors like diet and exercise are secondary. People's places in society—their level of income and education, the type of job they have, and their connections to others—are more primary influences on health.[47]

Although he is a fellow physician and a member of the research team for the Nurses Health Study, Kawachi's emphases differ profoundly from Walter Willett's. Kawachi's research shows, for example, that being poor or socially isolated increases a person's risk of heart disease as much as smoking. Being stuck in a job or relationship over which you have little control also significantly increases your probability of a heart attack.[48]

To focus primarily on people's personal habits is to miss the profound effect that community life has on health, Kawachi argues. His research shows that death and sickness rates from cancer, heart disease, and other major illnesses in the U.S. are higher in states where participation in civic life is low, racial prejudice is high, or a large gap exists between the incomes of the rich and poor and of women and men. By subtracting out the effects of smoking, diet, and other individual risk factors, Kawachi and his colleagues have been able to demonstrate that these social conditions influence health directly.[49]

Animal studies support their conclusion. In experiments where rabbits and monkeys are placed in isolation or in subordinate positions, or they are put under stress, their blood pressure and levels of "bad" cholesterol tend to increase.[50]

"We still labor under the myth that somehow we're each on our own and as individuals we can make these choices to prevent heart disease," Kawachi says. "If society or government really wanted to drive down the rates of cardiovascular disease, they would be tackling it at macro levels. Policies that appear to have little to do with health, like macroeconomic policies to reduce the level of income inequality, can have a major impact on driving down the rates of illness in society."

Narrowing the income and education gaps in American society would prevent disease and increase life expectancy not only among the poor, Kawachi contends, but throughout society. Don't count on health columnists, the American Heart Association, or the Center for Science in the Public Interest to redirect their spotlights in those directions, however. They are too busy taking the joy out of eating.

2

Safe Treyf

Pretending to Be a Saint

You have to eat something.

That reality can be a problem for disciples of the gospel of naught, whose advisers in the media and universities have renounced almost everything at the supermarket: meats, breads, and processed foods on account of their fats and carbs; fish and vegetables for the metals and chemicals they contain. So how do the faithful keep themselves alive?

Ultra-devout believers radically restrict their diets or raise their own food, and everyone else cheats. Like other orthodoxies, the gospel of naught has a small number of devotees who go to great lengths to comply with its teachings, and a much larger band of followers who find ways to convince themselves and others that they're faithful when they're not.

My favorite study of eating habits was undertaken in response to an oft-repeated joke in the Jewish community: "Why did the Jews starve for the first thousand years of our existence? Because, according to the Jewish calendar, the year is 5700-something, and according to the Chinese calendar, it's 4700-something. For a thousand years, Jews went without Chinese food."

Sociologists Gaye Tuchman and Harry Levine conducted interviews and dug up historical documents in search of an explanation for the immense popularity of Chinese food among New York Jews, dating back to the early 1900s. What they found is that Chinese food functions as what the sociologists call "safe treyf." Although some dishes contain treyf (nonkosher ingredients like pork and shrimp), these are minced and blended during cooking so they lose their distinctive taste and texture. And in line with another of the laws for keeping kosher, Chinese dishes do not mix milk and meat.[1]

Their interviewees told Tuchman and Levine that Chinese food is "close enough" to kosher that they could eat it without feeling guilty.

Skinny Pigs

I thought of that study as I munched on a pork sandwich in the exhibition hall at a convention of the International Association of Culinary Professionals. A young woman at the booth for the National Pork Board had handed me, along with the sandwich, several brochures on the nutritional benefits of pork, and a cool T-shirt. LET THEM EAT PORK was emblazoned on the front, and on the back, the organization's slogan: THE OTHER WHITE MEAT®.

The Pork Board's "other white meat" campaign is a brilliant marketing ruse to turn the ultimate nonkosher food into safe treyf. Dating back to the late 1980s, when preachers of the doctrine of naught were singling out red meat as particularly unhealthful, the "other white meat" campaign aims to get Americans to associate pork with chicken instead of beef.

Consumer surveys and increases in sales over the past couple of decades suggest the campaign has succeeded. It takes some doing, though, to produce barbecued pork as flavorless as what the Pork Board chose to showcase in my sandwich. This is the unfortunate price a food often pays in its conversion from treyf to safe. It loses its soul. Over the past two decades, to make their products seem more healthful, pork producers have removed much of what makes their meat flavorful. By changing how it breeds pigs and what it feeds them, the industry has reduced the fat content in pork by nearly a third. "Many cuts of pork are as lean as skinless chicken," it brags in its marketing materials.[2]

Skinless chicken breasts, of course, are the gold standard for carnivorous believers in the gospel of naught, and even the pork industry has capitalized on chicken's good name. In 2002, Hormel Foods, a company best known as the maker of Spam, extended its Always Tender line of marinated meats by adding chicken items. The previous year, Hormel had opened a 16,500-square-foot Spam Museum in Austin, Minnesota—to my way of thinking, a far more exciting development than all of their new flavored boneless chicken breast concoctions combined (Italian Style, Roast Flavored, Teriyaki, and Lemon Pepper). If I never see another boneless chicken breast, I'll die happy, but my life will be incomplete if I don't see the Spam Museum's 3,400-can Spam sculpture and seventeen-foot spatula flipping a five-foot Spamburger patty.

Jonathan Gold, the former restaurant reviewer for *Gourmet*, has written lovingly of the "porky essence" of Spam, "the over-generous nature of salty, fatty food manufactured for and revered by folks for whom salty, fatty foods is, or used to be, the ultimate in obtainable luxury. Spam is what this country is all about, a pig in every can and two cars in every garage." But

Gold's is a minority opinion, as is mine about boneless chicken breasts. Supermarket shoppers are so favorably predisposed to chicken breasts that companies like Hormel can market them as healthy even when they contain suspicious ingredients like "cheese flavor" and "partially hydrogenated cottonseed and soybean oil." Buried deep in the ingredient lists, those and other ingredients that many Americans consider unwholesome are what make Always Tender and kindred products palatable. As an article in a food industry trade journal put it: "The success of the Always Tender line is due to the proprietary 'Always Tender' injection formula, which includes a unique combination of potassium lactate, sodium phosphate and sodium diacetate to bind water after cooking and thereby retain moisture in the meat and impart succulence to the end product."[3]

Where the Flavor Comes From

Helpful hint number one for companies wishing to convert their products from treyf to safe: substitute chicken breast for the meat you currently use. Burger King's most successful new menu item in years is the Chicken Whopper. In just the first three months after its introduction in 2002, Burger King sold 50 million. Even the fast-food industry's most fervid opponent, Michael Jacobson, chief potentate at the Center for Science in the Public Interest, who frequently denounces items on Burger King's menu as "bigger and badder," blessed the product. The Chicken Whopper Jr. made it onto the CSPI's "Best Fast Foods" list. "Any grilled chicken sandwich makes a good meal. But, unlike some competing products, a Burger King Chicken Whopper actually tastes grilled," Jacobson's group declared.[4]

It *ought* to taste grilled. Burger King contracted with some of the world's top flavor chemists to contrive that taste. "There are three areas that contribute to what's between the buns," Peter Gibbons, senior director of product development at Burger King at the time and mastermind behind the Chicken Whopper, explained to me. "If I had to carve that up into a pie, I would give the flavor house 30 percent, the chicken processor another 30, and I would give the broiler at least 40."

The grilled taste of the Chicken Whopper results from running a processed chicken breast through a specially designed flame broiler that brings forward what the ingredient statement nebulously refers to as "Grill Flavor," "Smoke Flavor," and "Caramel Color." Those entities, whose formulas are closely guarded corporate secrets, are critical to the success of the sandwich. In industry parlance, the chicken itself is merely a "protein flavor carrier" that can be shaped, flavored, texturized, and colored in a limitless number of ways.

"If you were to ask me, can I taste the difference between unprocessed chicken from five different suppliers, I think the answer would have to be absolutely not. I don't think that anybody could," Gibbons told me. His principal criteria in selecting chicken processors are consistency and fulfillment; they must be able to provide hundreds of thousands of identical chunks of chicken as needed to satisfy market demand. For taste, Gibbons and other commercial food developers turn not to chicken suppliers, but to companies with names like Flavor Sciences Inc., International Flavors and Fragrances, and Heavenly Flavors. "Every time you develop something, you have a short list of whom you know you can count on and whom you can communicate with properly. We have four or five flavor houses we like to work with," Gibbons said. He didn't want to tell me which company he chose for the Chicken Whopper, but he did reveal what he required from it. The company had to be able to ensure that the sandwich would taste the same at Burger

King outlets worldwide. And the flavor profile of the product had to be uncomplicated. "We made a point," Gibbons told me, "of putting just two flavors in there. What you taste is very clean chicken flavor and the flame-broiling effect."

Because the Chicken Whopper is marketed as healthful to customers for whom, as Gibbons puts it, "health is a primary motivator," another restriction was put on the flavor chemists as well. "We wanted to have what we in the food business call 'a clean label.' We wanted consumers to have as much confidence as we felt it deserved, so we wanted it to be all natural."

Therein lies helpful hint number two for the marketing of a product as healthful. Call it "natural." When Americans see that word on a food package or in an advertisement, 86 percent of us assume the food is safe, according to a survey conducted by the National Consumers League. We consider anything labeled "natural" to be a pure and healthful gift bestowed by a benevolent Mother Nature. Tell us that something is "artificial," on the other hand, and we imagine toxic concoctions contrived by venal Dr. Strangeloves. That we think this way is understandable considering all the lethal chemicals, from nerve gas to DDT, created in laboratories over the past century, and the images of pristine landscapes in advertisements for natural foods. Those ads portray nature as gentle, clean, and restorative. The lethal aspects of nature—things like hurricanes, earthquakes, tornadoes, and the HIV virus—never appear in them.[5]

The absurdity of this conceit is not missed by people in the food business. Consumers may believe they are behaving virtuously by opting for foods with natural rather than artificial ingredients, but anyone close to the food business knows that the distinction requires, as Eric Schlosser put it in *Fast Food Nation*, "a flexible attitude toward the English language and a fair amount of irony."[6]

I had barely arrived at Flavurence Corporation and begun my interview with Dennis Beck, the company's president, when he rattled off an ominous list of toxins that sometimes find their way into foods. "Cheese listeriosis, salmonella, botulism, arsenic, cyanide," Beck said in what seemed to me a strange way to begin a conversation about a flavor company that bills itself as "the pure solution to a natural equation." My best guess was that he would go on to tell me he had built the company into one of the top flavor manufacturers on the West Coast by ensuring that none of those toxins could find their way into his products. Instead, Beck launched into a disquisition about how "unbelievably misguided" Americans are about natural versus artificial ingredients.

"There is a belief that if it is natural, it is safe, and if it's artificial, it is unsafe, but the real issue is healthy or unhealthy," he said. "We can all think of a number of natural products that are perfectly dangerous." Take crude oil, Beck suggested. "Perfectly natural. There is nothing more natural than that, and in its raw state, it's perfectly dangerous. Nobody is going to eat raw petroleum. But if we take this petroleum and distill it and start to burn it off, we get sugars." Flavor chemists use those sugars to make almond and strawberry flavors that are not only safe for human consumption, but so true to life, "they would knock your socks off," said Beck.

That those flavors have to be labeled artificial is ridiculous, he contends, since they're made from a substance that spews from the ground. And he offered other, equally curious examples, such as vanilla, which comes from an extract of the orchid bean; and vanillin, which he says tastes the same and is nearly identical chemically but is made from wood pulp. Create the taste of vanilla from wood, however, and you cannot call it natural. "So someone reading the label says, 'It's artificial vanilla.'

But it's not," Beck protested. "It's artificial in the sense that it didn't come from a vanilla bean, but it's not artificial in the sense that it's synthetically made." Wood pulp, he pointed out, "is as organic as they come."

Federal law mandates that for a flavoring to be called natural, at least 51 percent of it must come from what its name implies. This legal definition ensures that some natural flavors are actually *less* wholesome than the artificial versions, Beck argued. Among the many constituent parts of a fresh, organically grown strawberry are some that flavor houses are prohibited from using in their products because they are carcinogens. But because strawberry juice concentrates are used in natural strawberry flavorings, Beck explained, some of those carcinogens inevitably end up in those products. So does anything that the fruit had on it when it was crushed. "The strawberry may have pesticide residue, may have insect residue, can have fertilizer, can have E. coli from the fertilizer. Farmers do their best to wash it off, but they don't get everything off."

Listening to Beck, it was hard to imagine why any sensible shoppers would want actual strawberries in the flavoring for their strawberry ice cream. Not only might the fruit add contaminants, but it contributes almost nothing to how the ice cream actually tastes. The taste comes primarily from other ingredients in the "natural flavor"—the 49 percent that is not strawberry concentrate. Real strawberries allow a manufacturer to label the product natural, and consumers to feel less guilty for choosing it, but they dilute the flavor.

"Strawberries are 90 percent water, so what really happens is, you water down your good vanilla ice cream. You provide very little strawberry flavor to it, and now you've watered it down and your vanilla is not quite as good as it was before," Beck said, and to drive the point home, he showed me secret formulas for two of Flavurence's natural strawberry flavors and had me smell samples from vials of them. Each formula contained, in addition to strawberry juice concentrate, eight other ingredients: substances such as alcohol ethyl grain and furaneol. Derived from sugars, furaneol has the aroma of ripe strawberries. It is largely responsible for what I was about to experience, Beck told me as he placed a drop of liquid from the vial on a blotter and handed it to me.

When I held the blotter near my face, not only did I smell strawberry, I *tasted* strawberry. I had read estimates from chefs and food chemists that something like 90 percent of taste comes from smell, and now I realized why they said that. A little furaneol goes a long way. Six hours after I left Flavurence, I was still smell-tasting the flavor.

What "Natural" Means

That evening, while searching through Food and Drug Administration documents for information about some of the ingredients in Flavurence's formulas for its natural strawberry flavors, I saw firsthand how surreal the distinction between artificial and natural can be. Looking up ethyl alcohol in an FDA policy manual, I came upon this statement: "Practically and scientifically, pure ethyl alcohol synthesized from natural gas or petroleum products does not differ from that obtained by fermentation with subsequent distillation. Furthermore, foods in which one is used cannot be distinguished objectively from those in which the other is used."

So what reason does the government give for disallowing petroleum derivatives? "We believe that consumers generally expect the alcohol in food products to have been produced from fermented food substances, such as grains, fruit, etc., and that they do not expect their foods to contain 'alcohol' produced from petroleum gas."[7]

The "etc." in that sentence is important. Flavor houses may not be allowed to call a flavor natural if it comes from petroleum, but neither must they use only familiar fruits and grains, as most consumers probably assume. Were they so limited, they would be unable to produce many of their more complex natural flavors. Flavurence alone makes dozens of natural strawberry flavors. Some taste candied; others have a particularly fresh, ripe, or seedy taste; and they differ by their intended use. A strawberry flavoring that will end up in ice cream has different chemical properties than one designed to survive the high temperatures involved in making baked goods.

More complex still are flavors that imitate sauces or cooking methods. Product lists from flavor houses include beef gravy, Dijon mustard, hickory smoked, and slow roasted. Where the constituent elements for those flavorings can come from, and how they may be produced while still qualifying as natural, is spelled out in an FDA regulation whose breadth and peculiarity are best appreciated by reading it in full:

> *The term natural flavor or natural flavoring means the essential oil, oleoresin, essence or extractive, protein hydrolysate, distillate, or any product of roasting, heating or enzymolysis, which contains the flavoring constituents derived from a spice, fruit or fruit juice, vegetable or vegetable juice, edible yeast, herb, bark, bud, root, leaf or similar plant material, meat, seafood, poultry, eggs, dairy products, or fermentation products thereof, whose significant function in food is flavoring rather than nutritional.*[8]

In other words, to produce a flavor that qualifies as natural, a flavor company may ferment, fire, catalyze with enzymes, and commingle all manner of oddly unrelated substances, so long as those substances have been approved for use in food.

That's a far cry from what most of us imagine "natural" means. Surveys find that most consumers believe products with natural ingredients are simple and unprocessed, a misperception that the food industry routinely exploits. "The addition of a natural flavoring system allows the manufacturer to reposition itself into a premium market segment, resulting in higher margins than would otherwise be possible," the head of research and development for a large flavor house told a trade magazine. In the same article, the CEO of a consulting company that advises food companies recommended, "Even if there are loads of preservatives, stabilizers, emulsifiers and the like, if you are able to put 'natural flavor' on the label, then it can counteract the potentially negative effect of these other ingredients."[9]

Reading those bald assertions, it is tempting to hold the food industry entirely to blame for Americans' distorted notions of what qualifies as healthful. But there is more to the story. Although food makers have great sway over our minds and our purses, they do not work with blank slates. We are culturally predisposed to pitches for natural products. Paul Rozin, the food psychologist from Penn, brought me to understand this one day during a lunch he and I shared in New York. I had just taken a bite from our delicious Thai beef jerky at Rain, a restaurant that inventively updates Malaysian, Thai, and Vietnamese cuisines, when Rozin temporarily spoiled my appetite by describing one of his experiments. "If you drop a cockroach in juice, people won't drink the juice," he said. "Ask them why, and they say because cockroaches are disease carriers. So you say, 'Okay, we'll use a sterilized cockroach.' Makes no difference. They still won't drink the juice.

"When we confront these people on the fact that they don't

want the cockroach even though it's sterilized, they stumble or they're a little embarrassed. Eventually they say something like, 'It's a cockroach, I mean, it's inherently bad.' "

He found much the same, Rozin said, in a study where he questioned people about their views toward natural foods. "They say natural is healthier, so we say, 'Suppose something is chemically identical. Now you don't care that it's not natural, right?' The great majority of them say, 'No, I want the natural.' " Push people to defend their illogical view and they merely repeat that they think natural is better.

To tap into some of the less conscious assumptions behind this irrational thinking, Rozin developed a word association test that he administered to groups of people in the U.S. and France. By comparing the results, Rozin was able to identify a key difference between the two cultures. Americans tend to associate natural foods with health, while the French associate them with freshness. "There is not a sharp distinction between food and medicine in the United States as there is in France," Rozin explained. "You don't find drugstores selling food in France, and you don't find many drugs in food stores." This blurring together of food and medicine in American culture renders us susceptible to dubious reasoning about food.

Exercises in Excess

Consider so-called functional foods. Rather than double over with laughter at the sight of a candy bar that professes to boost the immune system, reduce stress, and burn fat, many Americans, some of them with college degrees, take out their wallets. All told, we spend about $20 billion a year on foods loaded up with vitamins and herbs said to ward off heart disease, cancer,

the symptoms of menopause, and practically every other afflic-
tion.[10]

"Eating cereal, I am told, will stave off depression," Elspeth
Probyn, a food scholar, writes in a recent book. "Washed down
with yogurt 'enhanced' with *acidophilus* and *bifidus* to give me
'friendly' bacteria that will fight against nasty *Heliobacter pylori*, I am assured that I will even lose weight by eating breakfast. It's all a bit much first thing in the morning when the
promise of a long life seems like a threat."[11]

A bit much indeed. Personally, when I see ads for fortified
breakfast foods, my mind flashes back to the priceless parody in
the 1969 cult film, *Putney Swope*. A black man, his wife, and their
two children are eating cereal at their kitchen table. In voice-
over, an announcer intones: "Jim Caranga of Watts, California,
is eating a bowl of Ethereal Cereal, the heavenly breakfast food.
Jim, did you know that Ethereal Cereal has twice as much vitamin B as any other leading cereal? Ethereal also has the added
punch of .002 ESP units of pectin."

To which Mr. Caranga responds, "No shit?"

By the time *Putney Swope* came out, ready-to-eat breakfast
cereals had already been marketed as tonics for nearly a century. Indeed, many of today's bestsellers are direct descendants
of products that William and John Kellogg created in the late
1800s and served at their sanatorium in Battle Creek, Michigan,
and those invented by their former patient, Charles W. Post. His
Grape-Nuts cereal was, according to Post's advertisements more
than a century ago, "the most scientific food in the world," a
"brain food" that contained "natural phosphate of potash . . . used
by the system in rebuilding and repairing the brain and nerve
centers."[12]

Are the promises on today's Grape-Nuts boxes to reduce the
risk of heart disease and make people energetic any less fatuous?

Given the regulatory hoops that present-day food companies must go through before the government allows them to make health claims for their products, and the endorsements on their packages from the American Heart Association, today's products certainly come off as more credible. But few of the 100 million Americans who buy functional foods to target specific health concerns get much benefit. "The increasingly common addition of vitamins and minerals to products as diverse as breakfast cereals, candy, and water is unlikely to provide additional increments in health and raises concerns about the possible hazards of too much of a good thing," reports Marion Nestle, a professor in the Department of Nutrition, Food Studies, and Public Health at New York University.[13]

No one doubts that some laudable forms of fortification have come about since C. W. Post's day. The addition of iodine to salt eliminated goiter, for instance, and milk fortified with vitamin D largely did away with rickets. But those twentieth-century additives were directed at deficiencies: they provided critical nutrients lacking in people's diets. By contrast, their misbegotten progeny, the functional foods of the present century, are exercises in excess.[14]

Take a product like Glacéau Revive, a water fortified with vitamin B, potassium, ginseng, and gotu kola. Although I was introduced to Glacéau Revive at a metacool press party in Hollywood where a twenty-something hunk was handing out samples, I am old enough to remember when Americans quenched their thirst with water straight from the faucet. Then came a much publicized study from the Environmental Protection Agency in 1975 showing that water systems in several major U.S. cities contained carcinogenic chemicals, and suddenly tap water came to be considered treyf. The bottled-water industry sprang up seemingly overnight. By 1980, sales of bottled water reached $575 million;

and by the end of the decade, Americans were spending more than $2 billion a year on the stuff.[15]

From there it was a short step to the cornucopia of fortified waters marketed today. If pure water is good, the logic goes, therapeutic water is even better. So now we have Dasani Nutri-Water from Coca-Cola, with vitamins the company says have been shown to improve metabolism, and Aquafina Calcium+ from Pepsi ("essential for bone health and strength"). Glacéau, the company that launched the enhanced-waters craze in 1998 with the release of SmartWater, offers more than a dozen specialized products, each with a different concoction of vitamin, mineral, and herb additives and named for the benefit it ostensibly provides—Focus, Defense, Endurance, Revive, etc.

Functional waters are definitely good for companies that sell them. Profit per bottle is typically 10 to 30 percent higher than for carbonated soft drinks, fruit-based products, or plain bottled water. How good they are for people who consume them is another matter. Because nutritional additives typically taste lousy, manufacturers use sweeteners or other flavorings as camouflage, which adds calories that many may not realize they are consuming. Glacéau Revive contains fructose as its second ingredient (right after water) and has 50 calories per eight-ounce serving. Drink the twenty-ounce bottle and you've downed 125 calories.[16]

Functional foods and beverages of the twenty-first century are actually a form of supersizing. The pitch for Glacéau Revive differs from what McDonald's used to do for its supersized fries only in audience. Both companies have implored people to pay extra for something they can do without. The McDonald's customer at least got what was promised. The functional-foods consumer gets only a promise. People who feel more mentally adept after drinking fortified water are probably responding either to the sugar in the product or to the placebo effect (having been told they will feel revived, they do). Even in the case of products whose additives have a legitimate scientific basis, the promised benefits can prove illusory.

Gold Circle Farms promises better vision, brain function, and cardiovascular health to those who buy its eggs enriched with DHA omega-3, a fatty acid that research has found to be beneficial for protecting against a range of diseases. But are fortified eggs the place to turn for omega-3s? A serving of sardines, anchovies, herring, or salmon has several times as much DHA. Moreover, the assumption that *adding* an ingredient to a food will yield the benefits that come from eating foods that *already* contain the ingredient is a form of reductionism. Many constituent parts make up a fish that is high in DHA, and they interact in intricate ways. Critics of functional foods point out that it may be those chemical interactions, rather than an individual nutrient, that matter for human health. DHA may have a synergy with other materials in fish that it does not have with the digestive tract of a chicken or the chemical components of an egg. (The DHA in Gold Circle Farms eggs comes from a marine algae fed to hens.)[17]

The premise behind functional foods is naive and potentially dangerous. Adding nutrients to our diets, indiscriminately, produces unanticipated consequences. An experiment at Columbia University showed, for example, that fortified foods can interfere with the ability of prescription medications to treat illnesses. Levels of the antibiotic Cipro were 41 percent lower in the blood of people who had been given the drug with a glass of calcium-fortified orange juice rather than with plain water.[18]

That study and others find that minerals in functional foods interfere with the body's absorption of Cipro and related antibiotics used to treat bronchitis, pneumonia, gonorrhea, and other illnesses. In principle, doctors and pharmacists can head off the problem by advising their patients not to take these drugs with vitamin and mineral supplements or particular fortified foods. But in reality, foods in this country are so awash in nutritional additives that a warning on the side of a pill bottle is unlikely to be effective. Millions of Americans down a bowl or two of fortified cereal and a fortified fruit drink—"the hottest trend in juice products," according to a trade magazine—and then consume more nutritional additives throughout the day.[19]

Men's Bread, Women's Bread

If forced to choose between the drugs their doctors prescribe for them and the functional foods they prescribe for themselves, many people probably would keep the foods. In surveys, three-quarters of Americans say they believe it is better to influence health through food than medicine.[20]

Aware of such statistics, makers of functional foods market their products as safe alternatives not only to conventional foods, but also to conventional medicine. They tap into what *American Demographics,* a magazine for the marketing industry, identified as "a trend spurring so many product categories these days: distrust of the HMO system." Several times a year, the functional-foods industry holds conferences where executives and product developers attend sessions with titles like "The Emerging Market for Disease Specific Foods and Supplements" and "Targeting Health Concerns with Functional Foods." Speakers at these gatherings use terms like "self-care consumer" and "food as medicine shopper" to describe people who are attracted to functional foods.[21]

With no apparent appreciation of their self-irony, the speakers at these conferences renounce modern medical science and conventional foods even as they evoke findings by medical researchers to justify health claims for functional ingredients. Also, many of the most popular functional foods are made and distributed directly by the world's largest food conglomerates, albeit under assumed names. Lest their functional foods be tainted by association with the treyf ones these companies make, they devise healthy-sounding names for them and promote them separately. How many people who buy Mother's Toasted Oat Bran Cereal or Propel Fitness Water realize they come from Pepsico? Or that Balance Bars and Boca Burgers come from Altria Group Inc., parent of Philip Morris, the tobacco company? (Neither the product packages nor the Web sites reveal the connection.)[22]

Other functional foods, while not made in their entirety by food conglomerates, include ingredients from them. In those instances, the connection may be particularly difficult to discern. When I picked up a sample of French Meadow Bakery's "Women's Bread with Soy Isoflavones" in the exhibit hall at a trade show for the natural-foods industry, I had no way to know that the featured ingredient came from Cargill, the nation's second largest food company. Only much later did I learn that fact. Neither the wrapper nor the full-color sales sheet for the bread indicated where the soy isoflavones came from. Both talked instead about unnamed scientists who say that isoflavones may prevent osteoporosis, heart disease, and cancer. And they claimed that other ingredients in Women's Bread counter the symptoms of PMS and menopause.

I was a little apprehensive about trying Women's Bread, but I am pleased to report that the two slices I ate did not cause me to grow breasts. Nor, sadly, did a couple of slices of the company's Men's Bread provide me with the bigger muscles, improved elimination, and increased libido the promotional materials promised.[23]

Perhaps to get those results, one must eat more than a few slices, something I would find difficult. French Meadow Bakery calls Men's Bread "megafood," "perfect food," and "smart fuel for your body." "What you taste is pure," it promises, but what I tasted was bitter and slightly medicinal. Nor did my luck improve when I tried samples of French Meadow's other functional breads, "Health Seed Spelt" and "Healthy Hemp." Directed at consumers who regard many of the ingredients in conventional breads as unhealthful, or who have medical conditions that restrict their diets, the products' wrappers boast they are oil free, sugar free, wheat free, dairy free, and yeast free. To my palate, they were pleasure free as well. They required too much chewing and had too many dissonant tastes to serve as a good vehicle for sandwiches or toppings.

There is no denying French Meadow Bakery's commercial success, however. The Minneapolis company sells more than 2 million loaves of functional bread a year in health food stores and supermarkets in every part of the country. The bakery is an exemplar of how a small but resourceful company can capitalize on long-standing beliefs about a particular foodstuff and, by updating them, create lucrative niche markets for itself. Ever since the first bakers discovered how to turn tiny, scentless grains into large, delectable, aromatic loaves, people have been crediting the staff of life with mystical properties (the body of Christ, manna from heaven) and the ability to stave off death and disease. "Bread has pride of place as a magic talisman against the evil eye, a vital substance," the late Piero Camporesi of the University of Bologna wrote in an essay about popular beliefs among Italian peasants in the preindustrial age. "It symbolizes the light of the sun and the great luminous spaces intimately linked with the power to fend off the forces of darkness, the underworld, death."[24]

French Meadow's product wrappers and promotional materials echo those sentiments, and after I returned home from the trade show, I arranged to meet the person who wrote them and creates the products. She is Lynn Gordon, described in French Meadow's press materials as "a certified macrobiotic cooking teacher by trade" and founder and president of the company. "My life's mission is advancing my own personal health and that of my family and my customers," she quotes herself in a press release. "I call it state-of-the-art bread with a consciousness," she says of her gender breads.

Nothing in the press kit prepared me for what I found when I visited the company's bakery plant and adjoining café a few months later. Visions of Hemp Bread stuffed with diced tofu danced through my head as I drove to French Meadow that morning. The names of nearby businesses in the south Minneapolis neighborhood where the bakery is located heightened my suspicions—Nature's Wisdom Health Shop, Clinical Nutrition Center, Ecopolitan 100% Organic Vegan and Raw Food Restaurant. By the time I walked into French Meadow's café, the last thing I expected was what actually awaited me: display cases packed with sweet, buttery tarts, tortes, scones, cakes, and muffins. The menu offered omelets with hash browns and bacon (the real stuff) and smoked turkey sandwiches on sourdough, rye, and (I kid you not) white bread.

Gazing in awe at customers' plates laden with foods I wanted to try, and thinking to myself it was little wonder this place had a line out the door while the culinarily correct eatery down the block was nearly empty, I barely took notice of the modest display of functional breads, off to one side. And when Lynn Gordon arrived, she led me immediately to the pastry case and asked me to help her select an enormous array of treats that a server brought to a table Gordon had reserved for our interview.

At first it struck me as supremely hypocritical of Gordon to sell in her café—and nibble at during our interview—foods that many consumers of her functional breads undoubtedly consider lethal. Indeed, at one point in our conversation, Gordon herself suggested that such foods cause cancer. "His illness is very typical of someone who eats very *ying*," she said of her father, who was diagnosed with lymphoma three years ago. "My dad was living on desserts and soda pop, so it didn't surprise me that he would get that cancer."

But in the course of our discussion of the ingredients in her Men's Bread and Women's Bread, I came to feel that Gordon's flexibility is probably a key to her success. It's one thing to talk the spiritual talk, but you don't get your products into mainstream outlets like Super Target, as Gordon has, unless you're willing to walk the corporate walk. When I asked how she decided what to put in the two breads, she explained that initially she had set out to include as many presumptively healthful ingredients for each gender as possible, but backed off after she was warned of possible lawsuits. A regulatory specialist from Cargill, the $50 billion global agricultural and industrial products company that supplies the soy isoflavones for her breads, urged her to remove saw palmetto and ginseng from her original recipe for Men's Bread. "You can't put them in a food," she quoted the Cargill person as saying.

At first, Gordon protested. Men need the ginseng for energy and the saw palmetto to protect their prostate, she argued. She objected also to Cargill's demand that she stop claiming that the fava beans in Men's Bread function as "a natural Viagra." The Cargill person had nearly fainted when she saw this on the wrapper, Gordon recalled. "She said that Viagra is a patented, trademarked name of a product that is owned by Pfizer. She said you have no right to take that name. They could sue you for everything you have."

Gordon countered that she had read an article saying that men who eat a cup and a half of fava beans get an erection. "I said if that's not a natural Viagra, I don't know what is," she recollected. But eventually she backed down on both points. "I'm sure she's right. She's paid a lot of money," Gordon told me. Besides, as far as the saw palmetto and ginseng go, the Men's Bread actually tastes better without them, Gordon said. "They do not have a nice flavor."

How About an Oat-Bran Beer with That Soy Burger?

The pop stars of the food world, functional ingredients catch on or not, stay hot or lose their luster, depending less on their intrinsic worth than on the level of support they receive from powerful corporate backers. Some, like saw palmetto, are sold principally by small manufacturers rather than major food companies and never make it big in spite of alleged health benefits. Other miracle ingredients enjoy huge success, then fade away for a time, only to have their careers resuscitated by companies with a vested interest.

Any American older than thirty probably can recall an example of this cycle: the oat-bran craze. After reports from the National Cancer Institute in the mid-1980s suggested a high-fiber diet may reduce the risks of cancer, newspaper headlines declared oat bran "Preventative Medicine" and "the Elixir of Living a Longer Life," and Kellogg's and Quaker Oats ran ads for their bran cereals that said much the same. In *The Eight-Week Cholesterol Cure,* a book that spent a year on the *New York Times* bestseller list, author Robert Kowalski claimed he saved his life with a low-fat diet that featured three oat-bran muffins a day.

By the end of the decade, Americans were shelling out more than $5 billion a year for oat-bran beer, oat-bran pasta, oat-bran snacks. ("The potato chip that's good . . . and good for you," read packaging for a product from Robert's American Gourmet Food, a company more famous recently as the maker of Pirate's Booty, the puffed-rice snack that *Good Housekeeping* magazine revealed contained more than three times as much fat as its label avowed. In response, a New York woman filed a $50 million class-action lawsuit against Robert's for "emotional distress and nutritional damage.")

The oat-bran frenzy came to an abrupt halt in 1990 when Harvard researchers reported that diets high in the stuff do not lower cholesterol levels any more than low-fiber diets. The scientists fed people one hundred grams of oat bran a day (five times the amount in a serving of Quaker Oats cereal), yet the subjects' cholesterol levels decreased only modestly, and no further than when the researchers fed them refined wheat. Now newspaper headlines proclaimed, "Oat Bran Bites the Dust" and "The Rise and Fallacies of Oat Bran." Sales of bran products took a nosedive, and a new expression, "oat-bran syndrome," entered the vernacular. To this day, commentators use it to refer to narrow scientific findings that have been blown out of proportion.

But Quaker did not give up on its star ingredient. The company sponsored research that refuted the Harvard study, and in 1996 it successfully petitioned the FDA to approve a health claim that can be used in advertising and packaging. "Oatmeal, a rich source of soluble fiber, has been scientifically proven to reduce cholesterol," Quaker's boxes and promotional materials announce.[25]

Therein lies the surest strategy of all for marketing a product as healthful: run a government-approved health claim on the box. Studies by market researchers show that Americans consider products bearing health claims purer and more desirable than equally nutritious products without such claims. And thanks to legislation and lawsuits engineered by the food and dietary-supplement industries, the FDA is largely powerless to reject petitions from food makers who selectively cite studies in support of their claims while ignoring studies that contradict them.[26]

Only large food producers have sufficient resources, however, to put together a successful petition and, once the claim is approved, defend it against negative findings by pesky scientists. Take soy isoflavone, the ingredient that brought Lynn Gordon and Cargill together and enriches several of the world's largest agribusiness conglomerates. Soy protein began to attract favorable press in the 1980s, when studies suggested it lowers cholesterol, but not until Du Pont successfully petitioned for a health claim in 1998 did soy hit the big time.

In the years following the FDA's approval, food companies introduced several hundred foods made with soy. Gardenburger, for instance, a smaller company that had been making soy burgers since 1984 and saw its sales increase 25 percent in the first two months after the health claim was approved, promptly added nonburger items to its line. Whether its Meatless Breakfast Sausage "reminds you of the homemade sausage you might find in a little family diner," as Gardenburger claims, is open to dispute, but there can be no doubt that the FDA ruling made it more palatable to shoppers. Likewise, Superior Healthy Cup Mocha Latte from Sara Lee Corporation, a company best known for frozen cheesecakes, may or may not "speak directly to the needs of health conscious consumers," as a Sara

Lee spokesperson told a trade magazine, but she is certainly correct that the product "is on trend with the tremendous growth and consumer interest in soy-based foods and beverages." By 2005, Americans were shelling out more than $4 billion a year for soy foods.[27]

The torrent of advertisements and PR campaigns propounding the benefits of soy easily drowns out the voices of knowledgeable scientists who raise concerns. Soon after the FDA approved the health claim, two senior scientists at the agency let it be known that they had written an internal memorandum opposing the action. Some journalists reported their doubts—studies indicating that soy consumption can increase the probability of thyroid dysfunction and women's risk of breast cancer. But by and large, the news media ignored those negative findings, and subsequent studies suggesting that men who regularly consume soy in midlife have greater brain aging in later years, and that while soy reduces cholesterol levels in some people, it may actually increase the risk of heart disease for others.[28]

Guided by the soy industry, reporters tend to cover research that casts the bean as a miracle food. No favorable finding is too small to warrant coverage. Reporters made much, for example, of a study in 2002 suggesting that hamburgers can be made more wholesome by the addition of soy. That study consisted of feeding regular ground beef to seventeen college students every day at lunch for a month, and ground beef supplemented with phytosterols, a component of soy, to an equal number. Cholesterol levels in the soy group fell significantly, according to the researchers, whose funding came in part from the soy producer ConAgra.[29]

Larger and more deserving of attention, but largely ignored in the media, was a series of studies that call into question the very gist of the FDA health claim, which asserts that consuming twenty-five grams of soy protein a day lowers cholesterol and reduces the likelihood of heart disease. In one study, Tufts University researchers put forty-two people on diets that included more than fifty grams of soy protein per day for six weeks, only to discover that most experienced little or no reduction in cholesterol. The Tufts researchers also went back and looked at the analysis upon which the FDA based its health claim. The favorable evidence came, they discovered, from a single laboratory and was not supported by other studies.

Isoflavones, the soy component zealously touted by Du Pont, Archer Daniels Midland, and Cargill, fared particularly poorly in the Tufts experiments. While whole soy protein produced modest reductions in cholesterol, isoflavones had no effect on cholesterol. Nor did they reduce the number or intensity of hot flashes in menopausal women, even after three months of high doses.[30]

A yearlong Dutch study of 202 postmenopausal women found that isoflavones improved neither cholesterol levels, cognitive function, nor bone mineral density.[31]

Soy isoflavones may actually be harmful to human health. When Marion Burros of the *New York Times* called eighteen scientists with expertise on nutrition and disease, she could not find one who was willing to declare that taking isoflavones is risk free. Some suggested the opposite. And in an essay he wrote after reviewing studies that point to possible dangers from soy, Sir Colin Berry, a professor of anatomy at the Royal London Hospital, concluded, "We have here all of the ingredients for a food scare—hormonal effects, what is usually interpreted by the press as evidence of carcinogenicity, and data suggesting teratogenicity [damage to fetuses] in humans."[32]

Even some soy enthusiasts in the scientific community have raised doubts about isoflavone-laced foods. Though they contend that the fifteen to thirty milligrams of isoflavones in an average Japanese diet help explain lower rates of heart disease and cancer in Japan, they worry that some Americans are ingesting three to ten times that much from soy milk, soy cereals, breads, energy bars, and supplements. Some worry too about claims by respected environmental organizations that rapid growth in soy farming in Brazil (the world's second largest producer, just behind the U.S.) is contributing significantly to deforestation in the Amazon rain forest.[33]

For all that, contrary evidence is no match for the $2-billion-a-year soy industry, whose agribusiness giants protect the reputation of their revered ingredient. They ply food and health writers and editors with free trips to farms, factories, and research labs, and meals prepared by prominent chefs; and each year since 1998, the large soy producers have sponsored a pair of conferences that showcase university researchers who receive funding from the industry, presenting papers with titles like "The Positive Effects of Soy on Prostate Cancer" and "Soy Isoflavones and Hot Flash Relief." One of the conferences also includes presentations by industry leaders on topics such as "Marketing to Health Professionals" and "Turning Miracles of Science into Meatless Meals."[34]

Dip Your Blueberries in Chocolate

Because complex chemical entities have multiple effects in complex systems such as the human body, and foods are complex chemical entities, well-founded arguments can be made that just about anything edible ought to be shunned—or, alternatively, consumed in large quantities to prevent disease. If that sounds like an overstatement, consider four very different ingredients: blueberries, chocolate, and the preservatives BHT and BHA. Only the first of these is considered a miracle food, and for that reason is included in many baked goods and cereals. The others are widely regarded as harmful, when in point of fact, all four contain the same putatively healthful substance.

Over the past several years, growers' groups have sponsored research suggesting that blueberries protect against all sorts of ills, from heart disease and cancer to macular degeneration. The Wild Blueberry Association of North America runs ads in trade magazines proclaiming that its product is "just what today's health-consumers are looking for . . . the perfect choice for exciting new product ideas." In fact, if your goal is to increase your intake of antioxidants, you might as well look for products that have preservatives. The blueberry industry may have trademarked its product as "Nature's #1 Antioxidant Fruit™," but BHT and BHA, two of the most common preservatives in packaged foods, can go it one better. BHT and BHA *are* antioxidants. Food makers add them because they prevent damage by oxygen.[35]

Ever since the 1980s, when studies associated them with cancer in rats, BHT (butylated hydroxytoluene) and BHA (butylated hydroxyanisole) have been considered ultra-treyf by food activists. But more recent research shows that if BHT or BHA causes cancer, it is only at doses far higher than used in foods. At the levels at which these preservatives appear in foods, they may actually help prevent cancer; at low levels, studies conclude, they detoxify carcinogens.[36]

Somehow, though, I doubt that products will don health claims on account of their BHT and BHA. The only substance I can imagine moving all the way from condemned to sanctified is chocolate, and it has a long way to go. Distrust of chocolate dates back at least to the Aztecs, whose legends tell of people becoming weak and prematurely old from drinking it, and more recently it has had a bad rap on account of its high fat content. A one-and-a-half-ounce chocolate bar contains half the recommended daily allowance of saturated fat.[37]

The vilification is largely undeserved, however. Chocolate does not raise cholesterol levels; on the contrary, it increases blood levels for HDL ("good" cholesterol) and supplies protein, calcium, and antioxidants. Indeed, as Norman Hollenberg, a professor of medicine at Harvard, has commented, "the evidence suggesting a health benefit of cocoa and chocolate is at least comparable to the level of evidence supporting positive health effects of green tea and red wine."[38]

People who eat candy live longer than those who abstain, studies by university researchers show. Journalists and preachers of the gospel of naught make fun of those studies partly because some have been funded in part by Mars Inc., the folks who bring us Mars candy bars, M&Ms, and Milky Ways. In 2002, Mars patented a method of processing cocoa beans that yields chocolate with higher levels of cocoa polyphenols, which research suggests may bolster the immune system and protect against cancer, heart disease, hypertension, periodontal disease, and gingivitis.[39]

What ultimately qualifies chocolate as the supreme functional food, though, is the chemicals within it that make people happier and more sociable. Scientists are only beginning to understand how chocolate activates chemical reactions in the brain that make us feel more cheerful. One theory centers on the presence of cannabinoids, the active ingredient in marijuana. Don't bother trying to M&M your way to a high—you would have to eat a twenty-two-thousand-pound chocolate bar to get enough cannabinoids—but it takes only a small amount, researchers say, to improve people's moods.[40]

Jeffrey Steingarten, the celebrated food writer, exclaimed

upon learning that the fat in chocolate has almost no effect on LDL ("bad" cholesterol), "God's in his heaven; all's right with the world." Eating chocolate improves people's moods, Steingarten submits, because it tastes great. "Eating anything delicious stimulates the production in the brain of endorphins, a natural analogue to morphine. Put another way, people crave chocolate because it brings them intense doses of sensual and aesthetic pleasure. This," he avers, "cheers them up."[41]

It probably also accounts for why chocolate lovers live longer.

3

Promises of the Fathers

How the Food Industry Sells Its Wares

There is a limit to how much the food industry can sell us, even if we're willing to get fat. Unlike the exercise machines in the basement that I fully intend to dust off and start using, and the pants in the closet I am certain will fit again, food spoils. Eat it soon after you buy it, or the only place to put it is the trash.

The food industry prospers by persuading us to pay extra for what it refers to as "added value." Not just fortified waters and omega-3 eggs, but *all* food products are marketed this way. For processed foods, the added value is convenience. I might be able to buy the raw ingredients for less, but if I pick up a frozen pizza at the supermarket, I can just pop it in the oven and serve the family dinner.[1]

Whether I *ought* to consider that a plus is another matter. Critics of processed foods urge me to forgo convenience in favor of freshness, authenticity, and healthfulness, virtues that other sectors of the food industry claim for *their* products. "Why choose organic foods? You are contributing to the long-term health and well-being of your family by choosing certified organic foods," reads a typical ad from an organic-foods company.[2]

But when I looked into this assertion, I found that it, too, is open to dispute. Marketers of conventionally grown foods point out that independent scientists at places like the U.S. Department of Agriculture and the American Dietetic Association find organic foods are not always nutritionally superior to or safer than other foods. A body of research suggests that the worst that can be said about the levels of synthetic chemicals in conventional foods is that they may be high enough in some cases to put fetuses and some children at risk. The danger to adults appears to be negligible.[3]

Does that mean there is no reason to pay an extra buck for a carton of organic milk? Or conversely, given the criticisms of processed foods, should we hang our heads in shame if we serve our kids a frozen entrée? Examining the claims and counterclaims within the food industry, and between the industry and activist groups, a reasonable person could come to feel that nothing in the local supermarket—or natural-foods store, for that matter—is worth its sticker price. For a long time, I felt that way myself. Lately, though, I have come to believe the opposite. From my travels in diverse sectors of the food industry, I have grown to appreciate that most food products have true added value, though not necessarily the ones hyped in advertisements.

An Organic Lunch

I recall the exact hour when I came to appreciate the added value in organic foods. I was in the Anaheim Convention Center at a luncheon at the Natural Products Expo, the main trade show for suppliers and retailers of natural and organic foods, and I was not favorably predisposed. Everything I had seen at the show prior to this luncheon suggested that the $15 billion organic-foods industry rested on half-baked notions we baby boomers held in our youth about purifying our bodies, souls, and nations by getting closer to nature.

Personally, I never went so far as to move to a rural commune and live off the land, but during college in the 1970s, I did find a house near campus where the landlord let me plant a vegetable garden in the backyard. My bible was *Organic Gardening: How to Grow Healthy Vegetables, Fruits, and Flowers Using Nature's Own Methods*. A paperback whose plain cover featured a large black-and-white photograph of a scrawny lad in a veggie garden who could have been me back then, the book was written and published by J. I. Rodale, the man the *New York Times* called "the guru of the organic food cult."

Rodale, whose original name was Jerome Irving Cohen, was born in 1898 on New York's Lower East Side. He had been a cigar-smoking auditor for the Internal Revenue Service and an electrical-devices manufacturer prior to moving to Emmaus, Pennsylvania, in 1940 to try his hand at farming and publishing. A canny pitchman who built a multimillion-dollar empire, in his writings and frequent TV appearances (he died during an interview on *The Dick Cavett Show*) Rodale came off as the groovy grandpa we all wished we had. First published in 1955, his *Organic Gardening* had an old-fashioned feel that was part of its appeal to us aspiring counterculturalists. We liked to believe we were part of something "new, yet in reality age-old," as Rodale described organic gardening.[4]

Following his instructions, I began to toss eggshells and orange rinds onto a mulch pile, ordered a box of earthworms from one of the advertisers in a magazine Rodale published, and convinced myself that the broccoli I grew in my little garden in suburban Chicago were more nutritious as a result. At the time, I didn't question Rodale's claim that there was so much evidence of the superiority of organic foods for human health "it isn't even funny." Compared with the Zen macrobiotic diets and other mystical claptrap that some of my friends were into, Rodale's seemed downright scientific.

A half century after *Organic Gardening* came out, there continues to be plenty of hocus-pocus in the natural and organics business. Prior to the luncheon at the Natural Products Expo, my morning had begun at a keynote address by Mark Plotkin, an ethnobiologist who trails around with Amazonian shamans in search of frogs and insects whose innards might be turned into "natural medicines." He showed gorgeous slides of the rain forest and dispensed platitudes about the spiritual basis of healing, but when he explained how his "favorite fungus" cures AIDS, I headed to the exhibition hall, where wholesalers of every imaginable edible offered samples of their wares. I tasted bad organic pasta, bad organic wines, unduly chewy organic cookies, and several repellent soy-based drinks, also organic.

By the time I got to the luncheon, I expected more New Age blather and inferior food. Instead, the conversation was rewarding and so was the meal: mesclun salad with spicy roasted walnuts; a stew of wild mushrooms and parsnip; marinated filet mignon; and a pasta salad niçoise with chunks of turkey breast. I can still conjure up the taste of the deeply chocolate fondant with a vanilla and toasted almond ice cream they served for dessert.

The other people at my table were likable, engaging managers from Organic Valley, a cooperative begun by seven farmers in 1988 that has become the nation's leading supplier of organic dairy products, representing more than seven hundred farms throughout the U.S., and bringing in annual sales of $200 million. It was thanks to them—and my subsequent verification of their claims—that I changed my mind about organic foods. In separate conversations, the woman seated to my left and the man to my right spoke with obvious sincerity about the social and ethical commitments of their company. They explained to me how they protect their farmers from price fluctuations, a nightmare most dairy farmers face nearly every month that has forced several thousand to go out of business or sell their farms to large conglomerates.[5]

My lunchmates spoke with passion and intelligence about how their members' farms are healthier for people and other animals who live or work on them. Parents can play with their children without fear of exposing them to pesticide residues on their clothes or tracked into their homes. On organic farms, unlike most large agribusiness farms, the cows, hogs, and poultry get to wander outdoors, and they tend to live longer and with less pain.[6]

My Chat with the Top Food Cop

Supporting better circumstances for the men, women, children, and farm animals that produce our food—now there is true added value, a good reason to pay more for organic milk. Too bad the organic industry does not spotlight that in its advertising and PR instead of insisting its products are safer than conventional foods. So wedded is the industry to this line that it actually rejects a food-safety procedure that destroys E. coli, salmonella, listeria, and assorted other pathogens that studies find are as common in organic as in conventional foods.

Throughout the 1990s and the beginning of the present decade, when the U.S. Department of Agriculture was developing official criteria that must be met for food to be labeled organic, the organic industry campaigned for an explicit prohibition on that process, known as irradiation. The USDA complied, and now organic-food makers market their products as having added value because they have not been irradiated.

If anything, they have less value. Food irradiation has been deemed safe by the FDA, the Centers for Disease Control, the World Health Organization, and the American Medical Association. So strong is the evidence in favor of irradiation that the American Dietetic Association actively encourages its members to promote consumption of irradiated foods. Even Jane Brody has recommended irradiation as a safe and effective food-safety measure.[7]

Public apprehension about food irradiation results partly from the name of this new technology. "There is something creepily 1950s about the very word," writer James DeWan noted in the *Philadelphia Inquirer*. "Irradiation still conjures B-movie images of droning B-52s, iridescent aliens, and panic-stricken crowds fleeing invisible rays of poison. It was common knowledge that radiation didn't make things safe; radiation made things dead."[8]

Exposure to radiation *is* undesirable, but the food-irradiation process, which uses gamma rays, electrons, or X-rays to kill pathogens by breaking molecular bonds in their DNA, does not leave radiation in foods. What it leaves is the possibility of ordering rare hamburgers without fear of a subsequent visit to the emergency room—a fact that has prompted some restaurants to adopt a posture opposite to the organic industry and promote themselves as places to go for rare burgers because their meat is irradiated.[9]

I was first introduced to irradiated burgers in 2002 at the Food Safety Summit, an annual event in Washington, D.C., mounted by trade associations for the restaurant and food-processing industries. In the exhibition hall at lunchtime, a food-irradiation company plied us attendees with thick, tasty burgers, and the take-home message was not lost on me. "The meat was rare, something you never see anymore, especially at a public event," I wrote in my notes.

The rare meat was the first of a series of surprises that noon. A few minutes after I finished eating, I ran into Michael Jacobson, executive director of the Center for Science in the Public Interest (CSPI). An archenemy of the restaurant and packaged-foods industries, he was the last person I expected to find there. But as it happens, his office is only a mile away from the hotel. He decided to stop by on his lunch break, Jacobson told me, giving me a perfect opening to ask if he'd had an irradiated hamburger. His answer was not what I expected. "I don't eat meat, so I wouldn't. If it were an irradiated veggie burger I wouldn't care," he said, breaking ranks with his comrades not only in the organic movement, but in the consumer-advocacy movement as well. Public Citizen, the consumer organization founded by Ralph Nader and usually a close ally of the CSPI, has been a vocal opponent of the new technology. Citing decades-old studies while ignoring numerous others that contradict their claim, Public Citizen contends that irradiated foods cause cancer and other health problems.

Jacobson disagrees. "I think the health risks are pretty minor," he told me. His own reservations about irradiation have nothing to do with the safety of the method. "Irradiation shouldn't be used as a Band-Aid," he says rightly. "You shouldn't be allowed to have a dirty plant and then fix it at the end with radiation. You should have a clean plant and, hopefully, mitigate the need. But if irradiation is the only way to ensure safe food, then, yes, irradiate it. I'd rather have irradiated food than dead people." Instead of battling the irradiation industry, groups concerned about food safety should zero in on farms, slaughter-houses, and food-processing plants, he advises.

Unlike many people from the food industry, who can be counted upon to endorse any product or procedure that benefits their business and oppose whatever puts them at a competitive disadvantage, people like Jacobson often take positions that are idiosyncratic. They will tell you their sole concern is what's best for consumers, but in real life, their reactions may be more personal. Food activists sometimes seem to choose their crusades according to whether companies' claims for added value offend them. The food irradiation company's claims did not have that effect on Jacobson, but another company's definitely does. What he really wanted to talk with me about that afternoon was something called Quorn, a meat substitute that few Americans had ever heard about prior to the CSPI's campaign against it.

Quorn is precisely the sort of high-protein, low-fat, cholesterol-free food the CSPI typically touts. Indeed, just a few weeks before my chance encounter with Jacobson, the CSPI's official publication had declared Quorn "darn good-tasting" and named it a "Best Bite." So I was a little baffled to hear Jacobson complaining about having found it on the shelves of a food store he patronizes. "Is that what natural-food stores are coming to?" he railed.

Had Jacobson revised his opinion upon eating Quorn, I could understand. Eric Asimov, a food and wine critic for the *New York Times*, has written fairly of Quorn's mock meat products, "The cutlet is burdened with the overbearing flavor of garlic powder, while the patty aims for breezy and cool but tastes like powdered salad dressing." It was not about the taste that Jacobson had second thoughts, however. His revulsion against Quorn developed after examining its ingredients. The product "introduces thousands of novel proteins into the American diet with no testing for allergen sensitivity," he told me.

Determined to get Quorn off the market, Jacobson set up a Web site, QuornComplaints.com, to gather testimonials about its ill effects. Beside a full-color photo of the front of a package of Quorn Cutlets, he put a button visitors can click if they think Quorn made them sick. Below the button, in a hint of what a person might want to write, he quotes from someone who purportedly experienced "severe, sudden nausea followed by a few hours of violent vomiting after eating Quorn." At another page on the Web site ("Hear from the Victims") he posts the accounts that came in. Some are horrifying, but as proof of Jacobson's assertion that Quorn "has been proven to cause severe digestive reactions" in roughly one in ten people who eat it, these anecdotes bear the same relationship to true scientific evidence as Quorn Cutlets do to veal chops. They're pungent, but no one ought to mistake them for the real thing.

When Jacobson forwarded his stack of testimonials to the Food Standards Agency in Britain, a director of that government body responded by pointing out that with 13 million people having eaten Quorn items during the previous year alone, and with the product having been available for nearly two decades in Europe, "it is not really surprising that you have been able to find people who appear to be intolerant to it." One person in 146,000 to 200,000 reacts badly to Quorn, studies find. (By comparison, 1 person in 300 is intolerant of soy products.)

Vegetarian groups came to Quorn's defense as well. When writers for *Vegetarian Times* looked into the controversy, they concluded that Jacobson was off base. More than a hundred studies conducted in the 1970s and 1980s found very few adverse reactions, the magazine reported. In addition to human trials, they cited animal studies in which "generations of rats raised on it proved to be just as healthy as rats could be."[10]

Size Matters

Jacobson has another objection to Quorn that resonates with a common refrain of farmers. Quorn is an industrially produced import. What got him concerned initially, Jacobson told me, was his discovery that, contrary to the manufacturer's claim that it "comes from a small, unassuming member of the mushroom family," Quorn is mushroom free. "It's not mushrooms at all. It's made from fungus that's grown in a vat in England someplace," Jacobson fumed.[11]

For many organic farmers, opposition to processed and distantly produced foods is more than a philosophical stance. It is a business proposition. With Archer Daniels Midland, Tyson Foods, ConAgra, Frito-Lay, Coca-Cola, and most other argibusiness giants selling organic products, about the only ways family farmers can survive are by becoming part of a cooperative like Organic Valley and by marketing their products as locally grown and selling them at farmers markets and natural-foods stores. Many small farmers have had to stop using the word "organic" altogether because they cannot afford the fees and inspections required by the federal organic standards or the salary for an additional employee to fill out all the paperwork. These farmers have taken to marketing their goods with value-added descriptors such as "bird friendly" (grown without removing shade trees for nesting birds), or simply "locally grown."[12]

Michael Pollan, the author of *The Botany of Desire* and *The Omnivore's Dilemma,* has been a particularly impassioned advocate for these farmers. "Organic," Pollan has written, "is nothing if not a set of values (this is better than that), and to the extent that the future of those values is in the hands of companies that are finally indifferent to them, that future will be precarious." Upon learning that the organic milk he drank originated on a "factory farm" thousands of miles away and underwent a high-heat process called "ultrapasteurization" to survive its long travel, Pollan switched to unpasteurized milk from a nearby farmer who has nine Jersey cows. After he began encountering organic TV dinners in his supermarket, he stopped shopping there and undertook an investigation.

In small print in the ingredients list on one of the organic TV dinners, Pollan found such staples of mass-marketed foods as "natural grill flavor" and guar and xanthan gum. "The label assured me that most of these additives are organic, which they no doubt are, and yet they seem about as jarring to my conception of organic food as, say, a cigarette boat on Walden Pond," Pollan writes.[13]

What alarmed him all the more was that the company that makes the TV dinner, Cascadian Farm, is owned by General Mills, one of the world's largest food conglomerates. General Mills bought it from its founder, a man named Gene Kahn. Now a vice president of General Mills who drives a Lexus with vanity plates that read "ORGANIC," Kahn has come a long way from his original vision when he started the farm in 1971. He drove a beat-up VW Beetle back then, and like other hippies who abandoned city life to return to the land, Kahn had the lofty ambition, as a Cascadian Farms publicity piece puts it, "to change the world's food system."[14]

Pollan concedes that to an impressive degree, Kahn succeeded. He got organic foods into conventional supermarkets and onto dinner tables in conventional households, with the result that thousands of acres of land throughout the country have been converted from chemical-intensive farming to organic. But to Pollan's eye, this achievement is not as praiseworthy as it may sound. He accuses large organic food producers ("industrial organic," he calls them) of abusing their land through excessive tilling, their animals through excessive confinement, and the environment through the pollution released and the natural resources squandered when foods travel long distances before landing on a plate.[15]

To be worthy of the name "organic," Pollan contends, food suppliers need to do far more than forsake evil chemicals. They must strive for what he calls a "countercuisine." The converse of the TV dinner, countercuisine would consist of food raised on family farms or small cooperatives who return as much to the soil as they take from it, shipped only a short distance, minimally processed, and eaten by "socially conscious consumers devoted to the proposition of 'better food for a better planet.' "

This is the ultimate claim for added value: foods that benefit not merely those who consume them, but the whole world.

But how realistic is this vision? It depends on whom you ask. As Kahn said when I reached him by phone at Cascadian Farm's office near Seattle, "Most of us have grown up a bit, and while we love the good old days as outlined in Michael Pollan's articles, unlike Michael we don't think it's an either-or proposition: either buy food that's manufactured or go out and milk your own cow. I don't think that dichotomy is anything more than a romantic fantasy. We love people that milk their own cows. But we also believe that people principally buy food in supermarkets today and that there's room for both."

Only producers of mass-market organics are in a position to respond to the desires of everyday shoppers for convenient, consistent, affordable food, Kahn properly points out. Before his company was part of General Mills, he had been unable to deliver products that routinely have all of those attributes, and for Kahn, it is a three-decades-old dream come true. "I'm like a kid in a candy store," he told me as he described the sophisticated consumer-testing, product-formulation, manufacturing, marketing, and distribution capabilities that have opened up to him. The people who work with him in those departments at General Mills have impressed him both with their expertise and with their support for organic production. The stereotypes held by some in the organic-foods movement about employees of big agribusiness companies do not match his experiences. "I don't know of anyone at General Mills," Kahn told me, "who doesn't believe that reducing the use of chemicals is an admirable objective."

Where Kahn's colleagues at General Mills differ from organic activists is in their *reasons* for embracing that objective. Rather than abstract principles and politics, they are motivated by the market. "Consumers don't like chemicals, and General Mills is all about consumers," Kahn said. Consumers demand foods grown without chemicals, so that's what the company makes.

More precisely, what a substantial number of supermarket shoppers want is conventional products made of organic ingredients. The first thing General Mills asked Kahn to do after it bought his company was to develop a line of organic cereals that looked and tasted like General Mills' perennial bestsellers. Kahn takes great pride at having succeeded in that assignment, and in the process, having converted part of one of the nation's largest cereal-production plants to a pesticide-free environment. But as he talked about it, a question kept coming to my mind: Does the world really need an organic version of Honey Nut Cheerios?[16]

"The world needs organic products, including Honey Nut Cheerios, because these are the products that people eat in this country," Kahn said when I asked. "These are products that people desire, and in order to produce these products, we need to produce organic oats, and in doing so, we create opportunities for farmers to improve on farm profitability and at the same time, improve environmental performance."

It was a masterly reply. In one fell sentence, Kahn fused the credos of mass-market food makers and environmentalists. He did not entirely respond, though, to the deeper complaint of people like Pollan about the intrinsic worth of industrially produced foods. To nail the point, I asked Kahn directly about his product line that had incited Pollan in the first place—TV dinners. Some in the organic movement consider it almost immoral to bring such products to market.

"If they don't want to buy it, cool, then don't buy it," Kahn responded. "It's all about choice. I'd much rather eat an organic entrée than a conventional one if given a choice. Personally, I'd rather have someone cook me a great meal than eat a frozen entrée, but that's not how I always live, and sometimes it's very nice to have the convenience of a delicious, organic, frozen entrée."

This is Big Food's trump card in contests over who best serves the public. The industry provides the choices the vast majority of consumers want, and the minority who have other preferences are free to shop elsewhere. The industry extends this principle only so far, of course; General Mills asks consumers how much sugar they want in their cereal, not how much to pay the company's CEO or workers in its suppliers' sugarcane fields. And Big Food uses choice as a weapon in battles with its critics.

But captains of the food industry really do believe in providing people with more rather than fewer options. Even behind closed doors and while under attack by their critics, they take that approach. In a breakfast session for a few dozen restaurant executives prior to the opening of the National Restaurant Association's annual trade show, I was the only outside observer as a senior officer from Burger King warned his colleagues of more lawsuits and editorials accusing their industry of making people fat and ill. "I strongly urge that you provide options and choices for your customers," Chet England advised. "If on the one hand you have a nice, juicy, high-fat half-pound burger on the menu, which is a great consumer preference, you might think also of having a salad on the menu, or a broiled-chicken option, or some other type of product to give the consumer choices."

Why I Can't Bring Myself to Demonize Big Food

Before I better understood the inner workings of several large research and development operations in the food industry, I doubted the sincerity of Big Food's commitment to serving us what we want. Phrases like "king consumer" get tossed around a lot by food companies, but I suspected that often, food companies dream up new foods and create demand for them by spending vast sums on advertising. I didn't believe company spokespeople who insisted that, far from trying to shove anything down anyone's throat, at every stage in the development process, they seek out consumers' views about every aspect of their products.

As it turns out, they do exactly that, even for the most basic products. For example, McCormick and Company, the seasoning maker whose little bottles with red tops are in everyone's pantry, recently built a state-of-the-art research facility with fully equipped restaurant and home-style kitchens where it brings professional chefs and home cooks to advise on new seasoning mixes and flavor profiles. Planned by a host of consultants who included theater designers, lighting engineers, and panels of consumers, the building's testing rooms have 150-pound doors, odor-control units, and special lighting. Equipped with cameras, microphones, and computers loaded with a program called Compusense, the testing rooms are designed to automate the collection and analysis of consumers' and professional tasters' evaluations of how seasonings smell, taste, look, and feel, both when tasted on their own and in prepared dishes. And adjacent to the testing rooms are observation areas where McCormick staff members watch the goings-on through two-way mirrors.[17]

All this is in the service of providing grocery-store shoppers and food-service cooks with seasoning blends that accommodate the flavor profiles and preparation methods they prefer. In recent years, for example, home cooks have told McCormick that they want to prepare highly flavorful, restaurant-style dishes in thirty minutes or less. So the company developed a product line called "1-Step Seasonings" for popular main dishes. Pick up a bottle of the chicken stir-fry blend along with boneless breasts and vegetables, take out a skillet and soy sauce, and in twenty minutes you're ready to serve a dish as pleasing as the stir-fries at many respectable Asian American restaurants.

There are limits, of course, to the level of quality that packaged foods can achieve. I would never suggest they can compare to what a great chef or devoted home cook can produce. But the reason is not a relative lack of culinary sophistication among those who create commercial foods. Before I learned about commercial R & D operations, I imagined them populated by technocrats who would just as soon dine on a Weight Watchers Smart Ones Bistro Selection from the freezer case as at a preeminent French bistro. Again I was wrong. Many of the people who create processed foods know far more about fine cooking than most of us ever will, and many of their creations are delicious.

The Weight Watchers Bistro line, for example, is the brain-child of an outfit in San Francisco called the Center for Culinary Development, which employs eighty-five consulting food authorities, including preeminent Bay Area chefs such as Hubert Keller of Fleur de Lys and Craig Stoll of Delfina, and does its product testing on people they select from among the 12 million tourists who visit Fisherman's Wharf each year. Several of its creations are good enough to hold their own in blind tastings with dishes from reputable restaurants. (I particularly like the Smart Ones filet of beef in peppercorn mustard sauce.)[18]

Across the bay, at Mattson & Company, the nation's largest independent R & D firm, the culinary talent and research capabilities are equally impressive, if less glitzy. Mattson does not boast of star chefs, but when I spent the better part of a day at its twelve-thousand-square-foot facility in an office park near the San Francisco airport, I was bowled over by how many of the rank-and-file staff were foodies and expert chefs. Many had, in addition to training in food technology, degrees from top-ranked culinary institutes, and the bookshelves in their cubicles and lab areas were filled with cookbooks, restaurant guides, and culinary magazines.

Touring the place, I became increasingly hungry from the aromas of tomato sauce, grilled meats, and frying oil, and the sight of men and women carrying platters of steaming hot noodles, pizzas, and tasty-looking dishes I could not immediately identify. Fortunately, my tour guide, Samson Hsia, executive vice president of the company, let me stop at some of the work-stations for tastings. At one of those, what looked and smelled like ordinary French fries turned out to be flavored corn fries. A woman in a white lab coat had prepared several samples, each with a different level of cheese flavor, the goal of the day's work being to reformulate the product in response to comments from consumer tasters who had found the original prototype too intense. Addressing their concern without making the product dull was going to be no small achievement, I came to understand after trying a half dozen samples, only the most robust of which had much to commend them.

At our next stop, I was invited to try what looked and smelled like the perfect accompaniment to the fries—grilled, medium-rare buffalo burger. There were three patties on separate plates labeled "A," "B," and "C," and Hsia explained that they were identical except for their age, the purpose of the test being to determine after how many days of refrigeration the meat ceases to taste fresh and appealing.

I decided to leave this particular test to the professionals. So Hsia took me to an adjoining room where a half dozen men and women stood around a stove top, each holding a plate with pasta. Mattson makes a point of using real china and silverware in all its tastings, Hsia explained, except for products designed to be "finger foods," and the company employs full-time dish-washers. At Mattson, staffers consider it important, even during the preliminary lab work, that foods be experienced the way they would be by consumers in their own homes.

A man in a white lab coat handed me a fork and a plate of what looked and smelled a lot like the spaghetti Bolognese at Caffe Capri, a little Italian restaurant a few blocks from my house. But Caffe Capri's is rich and hearty, while Mattson's tasted bland and starchy and was short on sauce.

Fortunately, before anyone asked my opinion and I embarrassed myself, the manager for the project explained that this day's focus was the pasta. The final product would be a line of frozen vegetarian entrées, each consisting of a meat analog and sauce over pasta, but today he was concerned only to identify suppliers whose pasta products are indistinguishable when cooked, and hence interchangeable in large-scale production of the products. The samples were intentionally light on sauce, and none of the tasters found the pastas acceptable, I was relieved to learn.

Mattson develops products for Campbell's, ConAgra, General Mills, Kraft, Procter & Gamble, and just about every other major American food company, but during my tour of the facility, no one volunteered the names of any of the companies for whom the foods were being developed, and I knew not to ask. With new product introductions costing tens of millions of dollars each, and competing companies scrambling to respond to the same consumer trends at the same time, product-development groups stringently guard the confidentiality of anything they are researching. It is uncommon for a writer to get even the limited degree of access I was accorded at Mattson. As a favor to the president of one of the client companies, the staff let me look around, but on the condition that I would ask questions only about products that were already on store shelves.

About those products, Hsia was forthcoming. After we left the labs and office areas, he walked me through rooms used for consumer tastings and others with industrial-sized mixers, refrigerators, and freezers; specialized ovens for making pizzas; and dozens upon dozens of containers, each marked with a separate starch, emulsifier, sweetener, spice, or preservative. Finally we ended up in a conference room where the company has a sampling of its creations on display: Starbucks Frappuccino, Mrs. Fields Semi-Sweet Chocolate Chip Cookies, Marie Callender's Beef Pot Roast, three varieties of Jack Daniel's Grilling Sauce, Boca Rising Crust Pizza (two varieties, pepperoni and sausage, both meatless).

Surveying the products as Hsia and I took seats at the conference table, I was struck by the phenomenal number of foods available in the U.S. to satisfy every craving, a luxury the likes of which no other people has enjoyed, and not to be belittled. Some of those cravings may strike some of us as questionable—the mere sight of the Frappuccino bottle turns my stomach—but there is no denying the ingenuity of the R & D teams who dream them up. Even with seemingly uncomplicated products, the technical achievements are impressive. Proudly framed and on display in the conference room was a patent for fresh pineapple wedges. Prior to the process that Mattson developed in the late 1990s, fresh sliced pineapple was unavailable for shoppers who liked the fruit but lacked the skill or patience to handle the prickly exterior and who considered canned pineapple déclassé. Mattson's client, the Maui Pineapple Company, had experimented with slicing pineapples and packaging them in individual plastic bags shipped by air to the mainland for quick sale in food stores. But the product's short shelf life coupled with the high shipping and handling costs made it unprofitable. And because Hawaiian pineapples are sweet only half the year, the product was unavailable the rest of the time.[19]

Mattson achieved something seemingly impossible. It found a way to package fresh, sweet-tasting pineapple pieces, year-round, that can be shipped by boat to the mainland and still have a week or more of shelf life. All told, Hsia and his colleagues spent a year experimenting with ways to add or remove the fruit's juice or quick-freeze and thaw the product, and they brought in panels of professional and consumer tasters to compare samples of their test products with canned and freshly sliced pineapple.

In the procedure they eventually patented, the sugar concentration (or "Brix score") and sugar/acid ratio of freshly cut fruit is measured, the fruit is chilled to near freezing, and juice is added. Throughout the year, Maui Pineapple pasteurizes and stores juice from pineapple crops with different levels of sweetness. Then for each packaging, those juices are blended to bring current fruit to the level of sweetness and acidity of ripe, in-season pineapple. During the winter, sweet summer juice is used to offset the tart taste of winter fruit, the result being, from the end user's point of view, delicious fresh pineapple in its own juices.[20]

Trouble in Paradise

Because Americans will pay extra for foods labeled "fresh," companies work hard to get that word attached to their products even when key ingredients come from storerooms. Whether they get away with it depends sometimes on whether their competitors successfully challenge them. The market for fresh pineapple being small, Maui Pineapple had no problem, but other food makers have not fared so well.

Paradise Tomato Kitchens Inc., a leading manufacturer of tomato sauces for restaurants, faced one of the more interesting challenges. In a formal complaint to the National Advertising Review Council (NARC), the ad industry's self-regulatory organization, Stanislaus Food Products, another major player, objected to Paradise's use of the word "fresh" in its advertisements for sauces made from tomato concentrate. Only tomatoes that have recently been picked qualify as fresh, according to Stanislaus, whose own products carry the tagline "packed from fresh tomatoes *not* from concentrate."[21]

To my mind, neither company's heat-treated, canned product seems to satisfy the FDA requirement that the term "fresh" be used only for "food that is raw, has never been frozen or heated, and contains no preservatives." Granted, the FDA goes on to cede a host of exceptions, for pasteurized milk, for example, and "fresh frozen" foods that undergo "brief scalding before freezing." But in its regulations, the FDA expressly forbids pasta makers from labeling their products "fresh" if they contain pasteurized ingredients.

The difference between the Stanislaus and Paradise products comes down to the number of steps involved in processing and canning. Stanislaus takes recently harvested tomatoes, crushes them, removes the peels, seeds, and water, adds seasonings and other ingredients, and heats and cans the resulting sauce. Paradise does much the same, but stores the tomato concentrate before combining it with fresh ingredients and canning.

Stanislaus argues that its sauce qualifies as fresh because it does everything at once; Paradise says its product actually tastes fresher because of its two-stage process. Nearly all tomatoes in commercial sauces come from California, which has a one-hundred-day growing season. If a company does all of its canning in season, as does Stanislaus, by the time its sauces end up in dishes, they may be a year old. With Paradise's method, on the other hand, the tomato bits may be just as old when they make their way onto a pizza, but the sauce itself—and some of its ingredients—is of more recent vintage.

Nonetheless, the review board ruled against Paradise. The word "fresh" implies that a product was prepared directly from newly harvested, unprocessed fruit, the board declared.

That did not stop the company, however, from continuing to include the word "fresh" in its marketing. Through clever use of a name it copyrighted for its two-step process, Paradise was able to comply with the ruling while actually making a stronger claim for its product. "Our proprietary All-Season Fresh Process© provides our customers with high quality sauces at the peak of their flavor profile throughout the year," the advertising avows.

If these sorts of wordplays and legalistic shenanigans seem absurd, so are the public's misconceptions that motivate food companies to sell their processed foods as fresh in the first place. Frozen and canned fruits and vegetables tend to be at least as nutritious as their fresh counterparts, but most food shoppers imagine otherwise. Consumers are largely unaware of contemporary techniques for flash-freezing and canning that retain micronutrients that are often lost during packaging and shipping of fresh produce. The levels of many vitamins decrease dramatically in fresh fruits and vegetables within several days after they have been harvested and refrigerated.[22]

Nor does fresh food necessarily taste better, as anyone can attest who compares flavorless fresh fruit, of the sort many supermarkets sell, with flavorful canned fruit. One of the few foods that are invariably better fresh is lettuce, and even there, food technologists have found ways to add salts and acids to the leaves so they can be broken off and still stay green and crisp in bagged salad mixes.[23]

In my conversations with food developers and chefs, frequently I was presented with counterexamples to the adage "Fresh is best." Pasta is probably the most familiar: long pastas such as spaghetti and fettuccini typically taste better if they have been dried. And the list goes on from there. Frozen peas are often sweeter than fresh, food industry folks like to point out, because the frozen are packed within hours after they have been picked, while the fresh lose some of their flavor during the several days it takes to get them to market and onto dinner plates. Dried herbs and chilis can be added earlier to dishes with long cooking times, thereby imparting richer and more even flavoring than their fresh cousins.

Then there are the many salsas, tomato sauces, and chutneys whose full flavor comes out only days or weeks after they have been prepared.

And how well do we distinguish between fresh and frozen anyway? The average customer at a sushi bar would probably be surprised to hear it, but most sushi in this country, regardless of how fresh it tastes, has been frozen at some point. FDA regulations designed to protect against parasites require that, with the exception of tuna, all fish intended to be eaten raw must be frozen first.[24]

Classy Dining

The hierarchy that elevates fresh over processed foods mirrors another: the social classes of the predominant eaters of each category of foods. Who is in a position to buy fresh foods and prepare meals from them? Those of us with time, money, and cooking and refrigeration equipment. People who lack some or all of these feed themselves and their families on processed foods.

It is an unappealing trait of fresh-food devotees that they envision purchasers of frozen and canned goods as half-witted dupes of the food industry. People go for convenience foods out of "ignorance borne of being fed a diet of peel-heat-and-eat," lamented a journalist in an online forum for food writers and scholars. His posting was one of dozens over the course of a week that deplored the demise of fresh, home-cooked meals and blamed it on the Beelzebubs at Big Food. Thankfully, someone eventually challenged the orthodoxy. "For my grandmother, the child of immigrants and surrogate parent to her four younger siblings while her parents worked for sweatshop wages," wrote Tracy Poe, a historian at DePaul University, "the invention of Duncan Hines cakemix and canned soup was a godsend that freed her up to decorate her post-WWII suburban tract home, have an independent social life, read (god forbid), travel, and play with her kids in the afternoon rather than slave over a stove.

"I got into cooking and food because I had the luxury of finding it fun and relaxing. But I still don't hesitate to order my family a pizza or feed my kids canned soup when I have had a long day at the office. I don't see it as a sign of cultural declension, but rather as a feminist victory that I can make this choice."[25]

The social status of fresh versus processed foods has shifted several times. In the latter decades of the nineteenth century, when canned foods first became widely available, they were expensive, high-status items that allowed the affluent to try fruits and vegetables unknown to their parents, such as pineapples and tomatoes. Then as the prices came down during the first two decades of the twentieth century, the wealthy drifted away from them, but canned goods became a symbol of successful Americanization for recently arrived immigrants and their children.[26]

By the 1950s, frozen and canned foods were popular at all levels of American society. A corporate lawyer living in suburban Chicago came home to roughly the same dinner as an insurance clerk in Levittown, New York. Both dined on what Harvey Levenstein, a distinguished food historian, refers to as "the All-American square meal," fare that would appall many of us today, but was revered back then: canned or dried soup, broiled meat with frozen French fries and a green vegetable (also frozen), and for dessert, Jell-O or store-bought ice cream.[27]

Some of the new commercial soups of this period—condensed cheese concoctions, for instance—were considered ingenious creations that enhanced any meal, whether served as a first course, a sauce for vegetables, or an ingredient in a casserole. Canned mushroom soups had been viewed the same way in the 1920s and 1930s, an era in which cookbook authors advised homemakers to improve the flavor of their homemade soups by adding a can of Campbell's.[28]

Lately, the reverse has become popular: home cooks are adding fresh ingredients to packaged foods. Several large foods companies are marketing products in which everything comes ready to heat and serve, save one or two ingredients the consumer buys fresh and adds to complete a main dish. The most thorough realization I have seen of this concept is a line of refrigerated dinner kits Kraft Foods introduced a few years ago called FreshPrep You Make It Fresh. Available in six varieties—lasagnas, alfredos, tacos, and enchiladas—the kits sell for about six dollars and include sauces or salsas, noodles or bread wraps, cheeses, and seasonings. A time-challenged parent need only add ground beef or chicken.

As a senior manager at Kraft explained to *New Products Magazine,* a trade publication for the prepared-foods industry, "Although moms don't want to spend time shopping or chopping, they do want some involvement with the meal preparation to feel they are providing a fresh and home-cooked meal for their families."[29]

Whether the meals Kraft devised to satisfy those desires have genuine value is a matter of personal judgment. Many moms evidently believe they do, but predictably, as soon as the products began arriving in stores, the Center for Science in the Public Interest vilified them. Rather than accept FreshPrep as one food option among many, none of them perfect, but all of which respond to genuine needs, the CSPI labeled the product "food porn." "Kraft gives you a quick, convenient way to serve fatty ground beef and three kinds of cheese to your loved ones," it inveighed, as if full flavor, speed, and tastiness were bad things that no respectable person would embrace.[30]

Most nights, most Americans lack the minutes, skills, or inclination to prepare the sorts of meals that the CSPI and other sermonizers would have them eat. That Big Food supplies them with what the industry aptly calls "meal solutions" is hardly condemnable.

4

Restaurant Heaven

Defining Culinary Greatness

I would compare the taste of this food to taking a flat stone and tossing it across a smooth surface of water so it skips. That's what this taste is doing. It's reverberating."

So reported Scott Haas, speaking almost breathlessly into his tape recorder midway through dinner at Daniel Boulud's eponymous restaurant, Daniel, on New York's Upper East Side. The chief psychologist at Arbour-HRI Hospital near Boston and occasional reporter for National Public Radio programs, Haas had driven four hours from Boston to New York to meet Boulud and experience his cooking. "The meal that I'm having here at this holy site, it's just so delicious, it's so tactile, it's so erotic, you just don't ever want it to stop," Haas rhapsodized.

Boulud's food provokes that kind of reaction in otherwise dispassionate people. When my wife and I dined there, by the time the meat courses arrived we were already in a state of rapture. And then a waiter placed before me a whole roasted squab, incomparably delicate and flavorful, served with crispy spinach and seared foie gras. For Betsy, the waitstaff brought chestnut-encrusted venison, braised red cabbage, sweet potato puree, and a cranberry compote.

This was the fifth movement of an exquisite symphony. The performance had begun with terrines of foie gras. Betsy's was served with cranberry-apple chutney and baby greens, mine with pheasant. Then came what I christened the Nouvelle French Sashimi Course: for the lady, a seviche of *hamachi* with red currants and a fennel custard topped with pistachio nuts; for me, tuna tartare with fresh wasabi, caviar, bits of cucumber and radish, and to cool it all off, a mild Meyer lemon coulis.

The Nantucket Bay scallops in our main fish courses, which came next, were sautéed to the precise nanosecond and then served with clementines and a cauliflower puree (hers) or wild mushrooms and bacon in a rosemary-infused lentil broth (mine).

A tasting menu prepared by a four-star chef like Boulud is to an ordinary dinner what a symphony performed by the New York Philharmonic is to a high school band practice. The second serves its purpose and may even be charming; the first is mind-blowing. The wonders that Big Food claims for its products and that nutrition nags allege for their diets may be far-fetched, but the splendor of a meal prepared by a great chef is real. Anyone who has the opportunity to experience such cooking should jump at the chance.

I don't say these things lightly. As I write the words, I am keenly aware that experts disagree about how to define culinary greatness. And a scene from the 1970s recounted in Ruth Reichl's autobiography comes forcibly to mind. The future editor of *Gourmet* magazine had just landed her dream job as restaurant critic for a regional magazine. She returned to the dilapidated house in Berkeley she shared with six fellow hippies and let them know the good news, only to have the bearded patriarch of the commune burst her bubble. "Let me get this straight," he responded. "You're going to spend your life telling spoiled, rich people where to eat too much obscene food?"[1]

The man had a point. High-end restaurants *are* places where the rich overeat in lavish surroundings, and the chefs who prepare the food are, in the words of Adam Gopnik of the *New Yorker*, "the last artists who still live in the daily presence of patronage."[2]

Those realities matter, not because they diminish the quality of the food (they do not), and not because they offend liberal sensibilities, but because they set culinary performances apart from other public arts. Anyone who can scrounge together the cost of a ticket can experience a performance by a celebrated conductor. The same cannot be said of a meal by a distinguished chef. Daniel Boulud would not have prepared this phenomenal feast for us had I made a reservation anonymously rather than through a friend who is a prominent food critic.

Given that we dined at Daniel during a busy period not long after a glowing review of the restaurant came out in the *New York Times*, I doubt we could so much as have gotten a reservation. It is one thing for me to propose that divine meals can be had here on earth, but quite another for everyone to have access to them—or even to know whom to believe about where to find them.

The Anonymity Myth

The reviewing methods of restaurant critics are supposed to preclude these un-American outcomes. To ensure that their experiences duplicate those of a nobody, they make every effort to reserve and dine anonymously.

The trouble is, often they don't succeed. Restaurant critics love to regale their readers and fellow writers with stories of odd names and hairstyles they've adopted; Ruth Reichl's wigs were legendary, and one critic I interviewed, Tom Sietsema of the *Washington Post,* has consulted with people from the CIA on his disguises. But chefs and restaurateurs tell a different tale. They say they post photos of reviewers in their kitchens and dissect their reviews so they can brief their staffs on the critics' predilections. Bob Kinkead, the chef-owner of several top-rated restaurants in the Washington, D.C., area, put it this way: "If you've invested $1.25 million in a project and certain people can crush you like a bug if they have a bad time, you're a fool if you're not trying to find out who these people are, what they look like, and when they're in."[3]

"The reviewers are so uptight about their anonymity and not being recognized, they take us for shmucks, like we're not smart enough to connect the dots," a prominent New York City restaurateur told me after I promised not to use his name. He said that he and his managers know not only what the major reviewers look like, but how they dress, the disguises they wear, the names they use on their credit cards, how they tend to order, and in some cases, whom they bring with them to dinner. On a couple of occasions, he reported, he has actually assigned people from his staff to sit at adjoining tables and monitor the critics' reactions throughout the meal.

"Critics here in New York are never anonymous, and they always write as though they are," Mitchell Davis, director of publications for the James Beard Foundation, told me. "I know firsthand, from people in every kitchen, that everyone knows who the critic for the *New York Times* is and when he or she is eating there.

"The best restaurants not only send out free food to the restaurant critic, they send out free food to everyone in that section, so the critic thinks everyone in the restaurant is always getting free food. It's sort of a joke," reports Davis, himself a former reviewer for the Mobil Guide.[4]

Anyone who followed the New York restaurant scene in 1993 probably remembers a much discussed episode that supports Davis's contention. In side-by-side reviews published soon after she became the restaurant critic for the *New York Times*, Ruth Reichl compared the insolent service and undistinguished food she received when she dined anonymously at Le Cirque, and the restaurant's outstanding performance at a subsequent dinner where she reserved under her own name. Though Reichl arrived a half hour early for the later visit, Sirio Maccioni, the restaurant's owner, let her know that "the King of Spain is waiting in the bar, but your table is ready."

Rating her first experience two stars and her second four stars, Reichl averaged the two and demoted Le Cirque from the four stars her predecessors had given the place.[5]

When I talked with William Grimes, Reichl's successor at the *New York Times*, he admitted "it's very hard" to be anonymous in top-tier restaurants. "I'm happy if I can get away with one anonymous visit, which I often can. Some places, the hawklike vigilance is such that it's almost impossible," Grimes conceded when I spoke with him over coffee in the cafeteria in the Times Building. (Critics from other leading newspapers and magazines took me along to restaurants they were reviewing, but Grimes was willing to talk only in the privacy of his workplace.)

He went to considerable lengths, Grimes emphasized, to

avoid being recognized. "I can't really go into it, but there are extremes to which I've gone that I never thought I would go, to ensure anonymity." And he argued that on those occasions when his cover was blown, the restaurants could do little to sway his review. "Broadway shows know when the critics are coming. Is every review a positive review because they know?" he asked rhetorically. "A lot of shows get completely trashed because they're no good. In the case of restaurants, they can know I'm coming, I can announce I'm coming, and they're still going to be crappy restaurants no matter what they do. In fact, often they shoot themselves in the foot because they assign too many waiters to the table and there's this constant filling up of the water glass and a kind of circuslike, ridiculous sending out of dishes."

Having experienced these circuses firsthand while dining with other restaurant critics, I don't deny his point. On one memorable occasion, a waiter was assigned to do nothing all night but ensure that the water glasses at our table remained full of an expensive imported bubbly water. When one of our party requested flat tap water instead, the poor guy was so flummoxed that all he could think to do was shuffle off to the kitchen for guidance.

But I've also experienced what Ruth Reichl described in her paired reviews of Le Cirque. At several top-rated restaurants in Los Angeles and New York, when I have dined first as a nobody and later with critics or big-spending customers the proprietors cared to impress, the contrast in the quality of the service and cuisine has been shocking. At one of L.A.'s best-known restaurants, where I've enjoyed some of the best meals of my life while dining with influential food writers, when I have dined as a nobody, I've been seated at tables in hallways and served lukewarm, overcooked food.

So I pushed the point with Grimes. I told him I knew from personal experience that restaurants deliver superior food and service to critics they recognize. "They can only be as good as they are," he answered back. "The chef can adjust a salad green so it's at the right angle on the plate, but he's not remaking a stock, he's not remaking a sauce, he's not buying the ingredients all over again. Ninety-plus percent of what he's doing is already set in stone. They don't know what I'm going to order."

This rejoinder, which I've heard from other critics as well, has morsels of truth. "You can't all of a sudden get better products in because he's walked through the door," acknowledged Tom Colicchio, with whom I happened to have an interview scheduled a couple of hours after I met with Grimes. A recipient of Best Chef awards and owner of multistarred restaurants in New York and Las Vegas, Colicchio went on to explain what he *can* do—and has done—when Grimes is in the house. He can make sure the waitstaff will give the critic attentive but not overbearing service, Colicchio said, and he can ensure that the food they bring to his table is the best in the place.

"I was here every time he was here, and the food was as good as it could possibly get," Colicchio said as we spoke in his office above Craft, his flagship restaurant. He described examining the salmon and lamb to select optimal cuts, inspecting the steak for gristle, and taking pains to season everything perfectly. "Your cooking is going to get better because the reviewer is in the restaurant," he let me know.

Yet restaurant reviewers remain determinedly blind to that fact. In awarding four stars to Thomas Keller's New York restaurant, Per Se, Grimes's successor, Frank Bruni, acknowledged he was "repeatedly recognized" during his several visits, and the staff "kept special watch over my table." Bruni wrote all of that off as largely irrelevant, however, even to his ability to evaluate the service. ("I kept watch over other tables and listened hard to acquaintances' reports of their experiences.")[6]

In addition to singular service, Bruni probably also got better food. If nothing else, the most accomplished hands in the kitchen worked on the dishes that landed at his table. In a book about the Herculean efforts Daniel Boulud undertook to get his restaurant elevated from three to four stars by Grimes, author and editor Leslie Brenner documented the care with which Boulud and Alex Lee, his executive chef at Daniel at that time, prepared the dishes that went to Grimes. "It's either naïve or expedient," writes Brenner, who spent much of the year 2000 in Daniel's kitchen, "to imagine that a cook with two months of experience can put together a shrimp, octopus, and squid salad with shaved fennel and lime-mustard dressing like Alex Lee, or that a line cook with fourteen months under his belt can do as good a job with a risotto or a skate wing as can Daniel Boulud with his genius and a lifetime of cooking. And unless you're someone very, very important, these two gentlemen will probably not cook your food.

"They'll keep an eye on what's going on—they'll taste sauces and make the quality-control rounds," Brenner allows, "but they will probably not touch your plate personally. If you do happen to find yourself among the lucky, lucky few who have the supreme good fortune to have one of them cook for you, the result will be something you will never forget."[7]

Important Personages

If my life depended on it, I could not recall the details of anything else my wife and I ate during the week that included our dinner at Daniel, but I can still call up the delicate richness of those scallops.

Chefs of Boulud's caliber intend nothing less. "Success equates to memories for me," said Thomas Keller, whose restaurants have been rated among the world's best in surveys of chefs and food critics. "The more memories you can give somebody, the more successful you are. It's not about money or about status, or any of that stuff, it's about actually touching somebody so much that they remember," he told me.[8]

But apart from those who write about chefs, who has the opportunity to secure those treasured memories? Who are the "lucky, lucky few," as Leslie Brenner puts it, for whom top chefs prepare meals?

In two words, the rich and powerful. Following my interview with Keller, I spent six hours observing in the kitchen at his Napa Valley restaurant, the French Laundry, during the evening dinner service. David Rockefeller was in the dining room that night, and I watched as Keller prepared a special tasting menu for him and his guests. I had eaten at the French Laundry on two previous occasions: once when I pulled strings to get in when Keller was away and a sous-chef was in charge, and once anonymously. Both meals were outstanding, but none of the dishes those nights compared to what Keller prepared for Rockefeller.

The chef let me try a couple of those ethereal dishes. The first, a whole foie gras poached in fig leaves, then smothered in figs and spices and oven roasted, Keller devoted a good part of an hour to preparing. The other, a glazed loin and shoulder of pork in a mustard seed and pork reduction sauce, served over savoy cabbage, Keller prepared along with a sous-chef.

Daniel Boulud goes further still. In the kitchen at Daniel, Leslie Brenner watched as he operated sixteen remote-controlled cameras concealed in the dining and lounge areas that allow him to keep an eye on guests. When he located VIPs, Boulud and his team prepared special dishes for them. His reservations staff also helped him keep track of VIPs by putting discreet codes beside their names in the reservation book.[9]

Managing that information—and much more about preferred customers—has become easier for places like Daniel and the French Laundry in recent years, thanks to a San Francisco–based company called OpenTable Inc., best known for its Web site, opentable.com, through which, in principle, anyone can reserve a table at any of more than three thousand restaurants across the U.S. OpenTable's appeal to restaurateurs like Boulud and Keller lies primarily in its data-tracking software rather than its reservations service. At sought-after restaurants, nearly all space at peak hours is held aside for VIPs. These restaurants tend to offer reservations at opentable.com for their earliest seating or during slow seasons. The French Laundry allots just two tables for each meal, but Keller and his staff make use of the software to other ends. They find it convenient for recording whether customers take smoking breaks between courses, which foods they like and dislike, and how they like these foods served. The level of detail is sometimes extreme: the entry for one woman says she likes only the inside of bread.[10]

Armed with such information from previous visits, Keller can time his cooking to allow intermissions for smokers, include lobster on the tasting menu of lobster lovers, and remove the crust from bread served to crust haters.

For the most important of their customers, chefs go further still. Wolfgang Puck tells me he is on the phone almost daily with chefs in other cities to arrange dinners at their restaurants for his VIP customers, or vice versa. The chefs share information not only about these people's favorite foods and wines, but about where they like to be seated, and things the service staff can do that will make them feel special.

So common is such fawning, a chef or restaurateur who opts for a more egalitarian approach can expect a tongue-lashing from the local aristocracy. Almost immediately after Los Angeles chef Suzanne Goin and her friend and business partner, Caroline Styne, opened their restaurant, A.O.C., food critics went into fits of ecstasy over the exquisite Mediterranean dishes and French charcuterie. S. Irene Virbila, the restaurant critic for the *Los Angeles Times,* declared A.O.C. "the star of this year's crop of new restaurants." At the *New York Times*, not only did Mark Bittman feature the place in an article on standout L.A. eateries, but Amanda Hesser flew out and did three installments of the paper's "The Chef" column on Goin.[11]

The upshot: throughout its first year, A.O.C. was one of the toughest reservations in town, and some of the city's royalty were in a snit when they couldn't get a table. "People have read things about other restaurants and have this idea that we all save tables for important people. The fact of the matter is, we don't save tables, we just fill the seats," Styne told me. "We don't care who they are or what their résumé is, and we get into a lot of trouble for this." When a big-time talent agent—a man who represents some of Hollywood's biggest stars—couldn't get a table, he had his staff find the name and home phone number of the owner. "He went insane," Styne recalls. "Calling me at home, having everyone he knew call me. 'I will never eat in your restaurant, ever, ever, ever. I will bad-mouth you.' "

With people like that, Styne believes, "there's a point where you have to say, 'I don't want you in here.' " But hers is decidedly a minority opinion among proprietors of fashionable restaurants, the most prosperous of whom owe their success as much to their skills at catering to the rich and famous as to the quality of the food they serve. Where would Wolfgang Puck be had he not ensured that movie stars, agents, producers, and studio executives were happy and ubiquitous at Spago in Beverly Hills? A genius at playing to the self-importance of players in the entertainment industry, Puck keeps them coming in, which in turn brings gawking tourists and locals who fill the massive three-hundred-seat establishment—and his other restaurants throughout the world—and view Puck as a star in his own right.

Making a Scene

These are not recent developments. Issues of access and privilege, democracy and exclusivity, artistry and artifice, emerged almost simultaneously with the birth of the restaurant in the late 1760s in Paris. Public eating places had existed well before this, of course, but they differed from restaurants in crucial ways. Taverns, inns, and boardinghouses fed travelers at tables d'hôte at fixed hours and for set prices. At restaurants, local people could order from written menus, dine anonymously, and eat what and when they wanted—differences that made a world of difference. Suddenly many people, not just the aristocracy, were able to become connoisseurs of food. In the words of historian Rebecca Spang, "within a restaurant, where every customer was presented with the same menu, social distinctions threatened to collapse into gastronomic equality."[12]

By the early 1800s, the *Almanach des gourmands* (*The Gourmand's Almanac*) was warning that those who would bring only money or titles to Paris's best restaurants, rather than a love of good food, could expect only indigestion. The *Almanac* and other accounts from the early decades of the nineteenth century describe a Parisian culture in which restaurants had become important social institutions, and eating was viewed as an artistic passion rather than mere biological necessity. Restaurateurs and chefs "were the equivalent of theater entrepreneurs and playwrights," notes Spang.[13]

The tradition of the restaurant *as* theater began during this period as well. On the one hand, cooks in Paris restaurants were honing and popularizing the cooking techniques, recipes, and overall aesthetic of modern haute French cuisine that would become the paragon in much of the world, including the United States, for many years to come. Yet at the same time, with their extravagant decor, the theatricality of their service, their fabled proprietors and patrons, and food that was sometimes distinguished and at other times merely showy, "as surely as restaurants relied on fish and fresh vegetables, silverware and champagne, they depended on legend," Spang documents.[14]

Some of the most successful restaurants of our day continue that tradition. On the opposite coast from Puck, Drew Nieporent has built an empire by wrangling celebrities to his restaurants not only as diners, but as investors. Robert De Niro, Francis Ford Coppola, Robin Williams, Christopher Walken, and Bill Murray have all invested in his restaurants over the two decades he has been in the business. Some of Nieporent's places—Montrachet, Nobu, Rubicon, Tribeca Grill—have served top-notch food, at least during some periods of their existence. But much of the time, Nieporent's empire thrives on buzz and exclusivity; people come to satisfy their egos as much as their palates.[15]

"It is like when you used to be able to get into Studio 54. It made you feel like you had status, you were part of an elite group," Nieporent told me over lunch at Tribeca Grill. Several of the most successful restaurateurs have maintained that, in *Los Angeles Magazine* food critic Patric Kuh's words, "the best way to get people in is to plant in their minds the idea they might not be able to."[16]

At our lunch, Nieporent ate a dish identified on the menu as "Drew's Salmon," a grilled fish on a salad of arugula and white beans, while I ate "Spice Rubbed Tuna," an entrée markedly more bland than its name implied, and listened to Nieporent explain what he referred to as "the seating order" at popular restaurants. "In New York, it's not what you eat, it's where you sit, and that is also a part of keeping a place hot, keeping it interesting," Nieporent said. "When you go to a movie theater for a premiere party, they have all of the good seats in the middle of the theater reserved for the cast members, friends of the cast members, the production people, the people who paid for the movie. And that's kind of what we do. People get preferential treatment based on the fact that they've accomplished something that's very high profile and celebrity."

Interviewing Nieporent is a scene in itself. His driver stands guard a few yards away like a Secret Service agent protecting the president. Our conversation is interrupted every few minutes by people from Nieporent's staff bringing him newspaper clippings about his restaurants or dragging him off to greet VIPs in another part of the room. On the wall are huge abstract paintings signed "De Niro '79." (The actor's father, Robert Senior, was an artist.)

Hanging around with Nieporent is not nearly as dizzying, however, as going along for a ride with one of his competitors, Jeffrey Chodorow, the wildly wealthy owner of Ono, Asia de Cuba, Spoon, and other assorted trendy restaurants in New York, Miami, Los Angeles, Las Vegas, and London. Chodorow is "as much a social director as he is a restaurateur," William Grimes once wrote. "Even when his restaurants don't have a bouncer and a velvet rope, they feel as if they should. The food, which tends to be a series of stylistic gestures, ranges in quality from so-so to dreadful," Grimes made known.[17]

For our interview, Chodorow asked me to meet him at noon at China Grill, the restaurant that he opened in 1987 in the CBS Building and that appears, year after year, on lists of America's top-grossing eateries. Grimes's admonition notwithstanding, I had looked forward to finally having a meal there, but when Chodorow met me in the lobby, he announced that our interview would take place en route to his current project. Then he directed me outside to Fifty-second Street, and the backseat of a black stretch limousine whose interior was big enough to host a meeting of the staff of a small company.

As his driver headed south, Chodorow laid out for me how he keeps his properties hot by hiring beautiful people as greeters and servers, locating in hip neighborhoods and hotels, and hiring public relations firms from the entertainment industry to schedule celebrity birthday parties, movie premieres, and fashion-industry events, thus ensuring that every supermodel gets photographed at one or another of his restaurants. And he constantly updates the ultrahip background music, menu designs, and servers' uniforms. "We start out trendy and we just never let it fade," Chodorow told me.

He insisted that serving great food is a key part of the equation as well. "I have this saying," he said; "it's not just about the food, but it's always about the food." Chodorow vehemently rejects the negative assessments by Grimes and other restaurant critics, but in my experience, their take is closer to the mark than his own. During our interview, Chodorow chose the word "fantastic" to describe the food at Asia de Cuba, but when I ate at the Los Angeles branch, I found it to be fusion at its dullest. Pretty people, yes, but pretty mediocre food.

Besides, if he honestly believes the food he purveys is so good, why has Chodorow been out hiring big-name, big-ego chefs? For years, he steered clear of those guys. "There's a big advantage to doing non–chef-driven restaurants, because you control everything. There are challenges to working with chefs because many of them are difficult," he told me. So what was he doing partnering with Alain Ducasse and Rocco DiSpirito for his newer restaurants?

When I asked him that, Chodorow admitted that part of his goal was to get prominent food critics to see him in a better light. If, by his association with acclaimed chefs, he can persuade them that he himself is serious about food, the critics will be more favorably disposed to all of his restaurants, Chodorow figures.[18]

Lights! Cameras! Bad Food!

Maybe he's right, given how Grimes reviewed the place where Chodorow's limo dropped us off that day, a short-lived restaurant on Twenty-second Street near Broadway called Rocco's that came about, Chodorow told me, as part of a deal he made with DiSpirito. When the restaurateur approached the chef to help him revamp the menu at another of his properties, DiSpirito said he'd agree if Chodorow would put up the several million he needed to open an Italian restaurant bearing his name and serving the kind of food his mother makes.

Familiar to fans of "reality TV," the pseudo-documentary genre that permeates prime-time television, Rocco's was the focus of NBC's show *The Restaurant,* whose second and final season ended with a face-off between Chodorow and DiSpirito for control of the operation. The moneyman eventually won; Chodorow's attorneys got a court order barring the chef from the premises. Initially, though, when the network went in search of a chef to feature in the series, DiSpirito, on *People* magazine's "Sexiest Men Alive" list and a regular on NBC's *Today* show, was an obvious choice, all the more since he was about to open a place of his own.[19]

The show continued a tradition dating back two centuries. "Paris's most famous restaurants were within the financial reach of only a tiny fraction of the population, but they were in the view and imagination of all," historian Rebecca Spang observed. Throughout the nineteenth century, the restaurants of Paris appeared frequently in novels, plays, and travel writing, where they were made out to be places of great intrigue.[20]

In our time, books by Anthony Bourdain and TV programs like *The Restaurant* have done the same. With Chodorow's blessing, I spent most of the afternoon and evening watching the TV crew film the cooks, waitstaff, and customers on the set . . . er, I mean, at the restaurant. With a dozen cameras mounted on the walls and ceilings, one-way mirrors with more cameras behind them, blinding lights, microphones all over the place, and assorted producers, directors, and camera and sound crews running around, it was hard to think of the place as a restaurant. This might have been easier had not nearly all of the waitstaff, bartenders, and chefs looked like models or actors, and several of them, when the cameras were pointed elsewhere, talked among themselves about their auditions and agents and how appearing on this show was helping their careers.[21]

Within an hour in the place one thing was clear to me: *The Restaurant* might be a reality show, but it was not going to show reality. The series may not have been scripted, but the staff performed for the cameras, and the people behind the cameras egged them on. I watched as producers told people where to stand and asked leading questions about the restaurant's customers, managers, and star chef.

I had little doubt which part of the day's action would end up on the air. The walkout by a waitress was so carefully choreographed, shot, and reshot that there was no chance it would end up on a cutting-room floor. All told, the show's producers filmed four exit scenes with this busty brunette. In one, she stormed out of the building and had words with DiSpirito; in another she picked up her belongings; in a third she cursed out a pair of managers. And for her final curtain call, a crew of five (two carrying cameras, two with microphones, and a producer) followed her to the entrance of the subway station. Before she descended, they had her sit for an interview in which they provoked her alternately to anger and tears.

When the series aired, this waitress was a focal character for a couple of episodes, and her departure got big play. The televised version included material from all four scenes, pieced together as one.

For all that, several seasoned journalists who should have known better failed to pick up on how contrived *The Restaurant* was. The show "opens a window that even professional food writers rarely get to look through," Grimes wrote in the *New York Times*. "The beauty of NBC's *The Restaurant*," Devin Gordon of *Newsweek* declared, "is that it's a glimpse of everything you never get to see: what your waiter says about you after he walks away, the chef dropping f-bombs in the kitchen, the mirror-fogging high jinks in the restroom."[22]

Reporters who bothered to poke around learned fairly quickly that the show was better characterized as "un-reality TV," as the president of the National Restaurant Association dubbed it in a letter of complaint to the network. A reporter at *USA Today* discovered, for example, that the show's producers handpicked the restaurant's staff as well as many of the diners who appeared on camera, and to create a cliffhanger for one of the episodes, they planted a restaurant review in the *New York Post*.[23]

When a writer from *Gourmet* interviewed one of the head

chefs, he was told outright that romances and conflicts among the staff had been instigated by the producers. DiSpirito himself admitted that some of the manipulations disgusted him.[24]

Gourmet let out another secret as well: Rocco's food was not ready for prime time. In contrast to William Grimes, who wrote in his "Diner's Journal" column that the food "made a strong first impression," *Gourmet*'s reviewer, Jay Cheshes, described the dishes as "better versions of what you might find at the Olive Garden."[25]

The critics' disparate opinions resulted, I suspect, from Cheshes's having eaten anonymously and Grimes's having been identified and served the best food. New on the job at the time, Cheshes was unlikely to be recognized. I cannot know with certainty, of course, that DiSpirito took special care with Grimes's meal, but the results of a little experiment I cooked up suggest as much. Before DiSpirito knew anything about me that would provoke special treatment, I ate at Rocco's and brought with me—unknown to the staff until late in the meal, when DiSpirito noticed her—someone he cared greatly to impress.

Selina Kayman, a food journalist who was a senior producer for NBC's *Today* show at that time, had eaten at Rocco's the previous week under her own name and loved the food. In an e-mail to me following that earlier meal, she raved about the "calamari fried lightly, very tender, served just with lemon—classic Italian preparation," and she praised several other dishes as well. "Good, honest, well-treated food," Kayman described it.

The meal we shared, much of which arrived cold, Kayman characterized as "a complete turnaround." "He was serving overcooked, tough calamari over marinara sauce. Why? Show me an Italian who eats calamari with marinara sauce," she said.

Defining Culinary Greatness

Food should be honest and authentic and made of first-rate ingredients handled with respect.

Do those principles provide the basis for a distinctive cuisine or culinary aesthetic? Theatrics aside, what defines gastronomic excellence in the present age? What does the cooking of today's most revered chefs have in common?

The old paradigm was clear: French haute cuisine. As recently as the 1970s, "best restaurants" lists for cities throughout the U.S. consisted almost entirely of places with French names, and the nation's culinary elite were dyed-in-the-wool Francophiles. The late Julia Child launched her career in 1961 with *Mastering the Art of French Cooking*, a book cowritten with two French collaborators; and for her legendary PBS series, she christened herself "the French Chef." Her buddy James Beard may be remembered as a promoter of American cookery, but he idealized French chefs. "Getting my first piece of French bread on the train yesterday made me realize again what masters the French are at the art. It seems to me," he wrote in 1955 during a trip to Europe, "that even the food on the wagon-lit restaurant was better than all the food of Italy."[26]

So categorical was the *New York Times*'s first restaurant critic on the matter—"great cuisine in the French tradition and elegant table service" are "time-honored symbols of the good life," the late Craig Claiborne declared in 1959 in a front-page *Times* article—that New York restaurateurs structured their menus and hired French cooks to accommodate his predilection. Should an elite restaurant dare to include among its offerings the food of a lower-status ethnic group, it could expect Claiborne's scorn. In 1960, in one of his earliest reviews, he lauded La Caravelle, newly opened on Fifty-fifth Street near Fifth

Avenue, for upholding "the best classic tradition." Claiborne pronounced La Caravelle "an establishment of such caliber, there is an inclination to use such expressions as 'first rank' and 'ne plus ultra.' " His sole criticism: at lunch one day, he was offered a dish better suited to an Irish refectory than a fine dining establishment. "There is a place for everything, but a restaurant of La Caravelle's genre and deserved prestige should not admit corned beef and cabbage to the menu even on a trial basis," Claiborne groused.[27]

Even Alice Waters, the woman credited with creating California cuisine, opted for the French name Chez Panisse for the restaurant she opened in 1971. To give the place a French vibe, Waters hired Frenchwomen as waitresses, and though her chefs were not French-trained, they and Waters strived to re-create the culinary tradition and ambience of a French country restaurant.[28]

Where Waters broke with the culinary ideals of the age and helped initiate a new creed was in her emphasis on fresh, seasonal, locally grown ingredients. That practice was commonplace in French country restaurants, but in America at the time, imported ingredients were viewed as superior. Now, thanks in large measure to Waters and her disciples, top chefs boast that they "seek out farmers, producers, and other artisans who take special care with their chosen craft, and can offer exceptional products just recently harvested," as a statement on the menu at a much-admired Chicago restaurant, North Pond, avows.[29]

To hear some of the culinary cognoscenti tell it, that statement pretty much defines the culinary aesthetic of our age. In this view, "the secret ingredient is ingredients," as Clark Wolf, a prominent restaurant consultant, put it to me. Today's best chefs distinguish themselves by using the finest ingredients that can be had, Wolf says, products that chefs of lesser talent or at second-tier restaurants either cannot afford or don't know when they see them.

According to Leslie Brenner, over the past couple of decades, "letting ingredients take center stage became the salient characteristic of American cooking." Rather than "just cooking French food in America," she says, chefs like Thomas Keller, Alice Waters, and David Bouley distinguished themselves and forged a new culinary ethic by finding the best ingredients and devising dishes that showed them off.[30]

There is no disputing the important role that artisan farmers play these days. Some have become famous in their own right. Chino Farm, a forty-five-acre operation located in an area of tony homes near San Diego, supplies Chez Panisse and Spago with fruits and vegetables, and has been profiled in the *New Yorker*. Lee Hefter, the executive chef at Spago Beverly Hills, buys from several dozen specialist farmers, many of whom, he says, are so obsessed with improving their products that they are lousy businesspeople. "I might have an order of quail and the quail guy says, 'I didn't kill them this week; it's too hot and they're panting.' But this is a passionate guy. And ultimately he produces the greatest quail," Hefter reports.[31]

At the French Laundry and Per Se, I have been served lamb so much more flavorful than at other restaurants that I have all but sworn off the meat elsewhere. Much of the credit goes to Thomas Keller's lamb supplier, Keith Martin, who feeds his flock on the highest-quality alfalfa and keeps them comfortable and antiobiotic free throughout their lives. For some of Keller's dishes, he goes further still. "I can say to Keith, 'I want you to feed your lamb ten pounds of salt today,' and he'll start to raise them," Keller told me, "and then we'll butcher some of them, taste them, and try to get the quality of lamb that we want, a certain flavor component. We want a lamb that has a certain saltiness to it, a natural saltiness. We want it to be a certain weight because we want a specific size that can complement the rest of the menu."[32]

When Keller seeks a particular mushroom flavor for a dish he's creating, he sends his mushroom supplier to the woods in search of a variety that will impart it. Sometimes his produce requirements have been so idiosyncratic that Keller has had to grow his own. For some potato dishes, he needs spuds that have been watered at such precise levels that he can't find a grower to supply them. "If we're going to make a puree of potatoes, we want a potato that has a really low moisture content so we can put more product into it. We can put more butter into it, we can put more cream into it, and the potato has a more intense potato flavor because it has less water, so there's nothing to dilute the flavor of the potato," Keller told me.

At one point in our conversation, Keller came right out and declared, "The most important thing you can do as a chef is get the highest-quality products you can get and treat them respectfully." But in point of fact, he's being a bit self-effacing here. The extraordinary food Keller and other great chefs prepare cannot be reduced solely to the ingredients they use. Provide the same materials to less talented cooks, and the results will be unimpressive.

The reverse is also true. A master can make magic with common materials. Itzhak Perlman may cherish his eighteenth-century Stradivarius, but he can make heavenly music on a borrowed violin. And master chefs can create remarkable dishes with ingredients anyone can buy. During the week of our interview, Keller had added to the tasting menu a salad that featured warm endive in a banana curry. The bananas he bought at a local supermarket. "We're not always looking for exotic things," he explained when I expressed surprise at that revelation. "For me, the idea was the texture of the banana and the flavor of the banana. I eat banana every day and I just enjoy the flavor of a properly, mildly ripened banana."

The issue, Keller said, is not getting the right banana, but handling it properly. "The banana becomes the focal point of preparing that dish: making sure that it's getting cut correctly, at the proper time, and not letting it sit for too long."

In reality, even chefs who tout the use of artisanal ingredients don't shy away from conventional products if those suit the needs of a dish. When I interviewed Bruce Sherman, chef and co-owner of North Pond, the restaurant whose menu boasts of "exceptional products" from artisan farmers, he acknowledged, "If side by side the small farmer's product and the industrial product are equal, I'll go with the farm product. If the industrial product tastes better, I'm going to choose the industrial product."

By "industrial product," Sherman means anything that comes from a so-called factory farm. But some distinguished chefs use *truly* industrial ingredients—mass-manufactured, brand-name products that would shock their customers if they knew. In the pantry at Daniel, alongside artisanal olive oils and rare lentils are boxes of Minute Rice and bottles of Heinz ketchup. At Cafe Atlantico, a top-rated restaurant in Washington, D.C., a dish described on the menu as "ceviche of minty mussels" contains pulverized Altoids ("the curiously strong mint"). The secret ingredient in the barbecued salmon at Equinox, another reputable restaurant in the capital, is Coca-Cola.[33]

A Restless Quest

Examples like those suggest an alternative vision of culinary artistry in our day. Perhaps it is not the freshness, uncommonness, or provenance of the ingredients that defines our culinary aesthetic. Maybe what distinguishes great contemporary cooking is identifying precisely the right ingredients for each dish, and the right way to prepare them.

That description seems to me closer to the mark, but it is important not to mistake this approach for perfectionism. In Michael Ruhlman's book *The Soul of a Chef,* the word "perfect" appears so often in the chapters on Thomas Keller that I lost count after a few pages ("the meat is perfectly seasoned, but perfectly, you couldn't improve on it, it is perfect . . . the vegetables, cooked perfectly"). Having observed Keller and fellow four-star chefs at work, I can appreciate the impulse. Just the care with which he adorns a plate with herb-infused oil screams of perfectionism. Like a miniaturist, he bends over his workstation to within a foot of the plate and squeezes from a small plastic bottle in his right hand a precise semicircle of tiny green beads.[34]

But top chefs reject the very idea of perfection. At the beginning of his *French Laundry Cookbook* Keller rightly declares that "there is no such thing as a perfect food, only the idea of it." The only imperative is that the finished product be delicious. In the first article she ever wrote about the French Laundry, Ruth Reichl ordained the place "as good as a restaurant gets in this country, maybe as good as a restaurant gets anywhere," but she didn't call it perfect. On the contrary, in that 1997 piece in the *New York Times,* she described Keller as someone who is "not afraid to make mistakes." What makes a meal at the French Laundry so special, she noted, is his ability to "draw you into his world, make you a participant in his restless search for flavor."

From my interviews with and observations of chefs and food critics, I have come to believe it is that search that sets apart the culinary ethos of our age. Whereas Craig Claiborne singled out in his review of La Caravelle the veal dishes in cream sauces prepared and presented in "the best classic tradition," in hers on the French Laundry, Reichl focused on "a dazzling quartet of contrasting flavors that arrive in espresso cups"—a bitter sorrel soup, a tomato consommé that is "crystal clear but tastes bright red," and a pair of soups that look similar but turn out to be totally different ("one is an ineffably rich lobster bisque, the other a clean, smooth puree of cranberry bean").

Different chefs perform different sorts of magic with flavors. Keller and his protégés amplify and surprise. "It's like building a house; you want to use different elements to construct it, but not necessarily do you want to see them. In cuisine you don't want to taste them on their own, but they want to be there to support the structure of the dish," Keller said. To illustrate the point, he told me the secret of a dish I had been served the previous night when I had dinner at his restaurant. To all appearances a simple chocolate sorbet, it was at once the lightest chocolate dish I think I have ever had, and among the most intense. The experience was unexpected: when I put the sorbet in my mouth, at first it tasted salty rather than sweet, but that sensation immediately gave way to a feeling of being in chocolate heaven.

"Salt makes your taste buds react to the primary component even more, so that it jumps out on your tongue," explained Keller. "One of the techniques that has to be mastered to be a really good cook is understanding seasonings and how they play with one another and play with the primary components. We're talking about a counterpoint. You want the salt to make the chocolate more intense. You want pepper to make the tuna a little more intense. You're heightening one flavor by adding another flavor to it."

Other celebrated chefs play with flavors in other ways and to other ends. For Tom Colicchio at Craft in New York, the point is to *avoid* adding other flavors to the principal ingredient. "My idea is, if you get great peas, that's all you ought to taste, great peas. That's what I want to taste. I don't want to taste peas with curry powder and this and that, just peas," Colicchio told me. "The flavor that we're going for is the flavor of the ingredient." He said his goal at Craft is for customers to tell him they never knew that peas could taste so great.

At North Pond in Chicago, Bruce Sherman has a somewhat different ambition. "I'm looking to turn people on to the flavor that's in foods they either haven't tried, or what's even more rewarding for me, they didn't realize they liked," said Sherman, whom *Food and Wine* magazine recently named one of the best new chefs in America. He served me a salmon dish in a chervil broth with wild herbs that made his point better than he probably realized. Since salmon is already popular with his customers, it was the accompanying vegetables—snap peas, leeks, and turnips—for which he hoped to bring a new appreciation. His goal, he explained, was to have customers come up to him and tell him they never liked those vegetables before they had them that night. But in my case, the reverse was true. I enjoy leeks and turnips and dislike salmon. Having been served pink slabs of the fish at umpteen dinner parties and charity events over the years, I can hardly bear to look at it.

Sherman's wild Alaskan salmon was a different story. Slow roasted rather than grilled or heated in a hot pan, it was so moist that I did a double take to verify it had been cooked through. And the fish actually tasted wild. If there is an equivalent to "gamy" for fish, that's how this salmon tasted.

Outreach

For all the variety in their approaches, today's best chefs share an assumption that sets them apart from their predecessors. They take for granted that their search for flavor should extend well beyond the United States and France.[35]

Before Bruce Sherman enrolled in L'École Supérieure de Cuisine Française in Paris, he spent nearly four years in India, a formative period in which he developed an appreciation for spices he had not previously known. In 2003, when Wolfgang Puck devised the menu for a dinner in New York hosted by the James Beard Foundation to celebrate Beard's hundredth birthday, he included no traditional French or American dishes. The meal began and ended with flavors from Puck's native Austria (a soup and salad made with white asparagus flown in from his favorite Austrian grower, and Kaiserschmarren, Puck's signature dessert soufflé). And the penultimate dish of the evening was roasted rack rack of lamb "chinois," served with Hunan eggplant, wild mushrooms, and a chili-mint vinaigrette.

At Spago Beverly Hills, Puck's menu routinely includes dishes such as "Thai-Style Chicken Salad" and "Tahitian Vanilla Angel Cake."

The best sashimi I've ever tasted was at Per Se, where Keller served me shad roe "porridge" topped with shaved bonito and a little lime salt.

Even traditionally trained French chefs search afar for flavor. Daniel Boulud can draw a direct line from his mentor, Michel Guerard, back to the most acclaimed and influential chef in French history, Antonin Careme, author of the 1833 treatise *L'Art de la cuisine française*. But in his own essay published in 2003, *Letters to a Young Chef*, Boulud praises Gray Kunz, the former chef at Lespinasse, for "his balance of the Indian, French and Chinese palates" in dishes like mussels in lemongrass broth, and walleye pike with lavender honey sauce. Boulud encourages aspiring cooks to study the flavors of the cuisines of Spain, Italy, India, Mexico, China, and Morocco.

For his own success in dishes like the tuna tartare he served during my meal at Daniel, Boulud credits late-night meals at sushi bars in New York and a visit to Peru. With wasabi, caviar, cucumber, radish, and a Meyer lemon puree, that dish ingeniously unified flavors from Asian, South American, and French cuisines.[36]

By integrating the flavors of previously undervalued or unfamiliar cuisines into the food they serve at their restaurants, chefs of Boulud's and Puck's stature have helped to legitimize those cuisines in their own right. Until recently, Americans were unlikely to explore the great cuisines of the world unless they lived in ethnic enclaves. And even then, people tended to have access only to the food of their own ethnic group, or to bastardized dishes like chop suey.

Today, most every major city in the U.S. boasts dozens of serious ethnic eateries; New York and Los Angeles are home to hundreds. Many serve primarily their own ethnic communities, but a substantial number live off of a group of restaurant goers on a flavor quest of their own. Who these people are and the debates that have raged over their lifestyle and what that tells us about American society is the subject of the next chapter.

5

The Food Adventurers

In Search of Authenticity

The best cooking of the nation's best chefs may be available only to the rich and famous, but anyone who is open to trying a range of ethnic cuisines can eat out well in America, and for little money.

Finding those eateries is not as simple as it may sound, however, in a nation where eating carries so much moral baggage and the varieties of food snobbery are numberless. Almost anyone you might consult for a restaurant recommendation has had his palate addled by one ideology or another. Either he embraces the gospel of naught that damns Latin American food for its fat content or Asian food because rice and noodles are high in carbs, or, out to prove his cosmopolitanism, he extols only the most arcane dishes served at the most out-of-the-way places: pig offal soup at a Thai dive in Queens; fish head stew in a Chinese restaurant an hour's drive from San Francisco.[1]

So whom can you trust? In my experience, about the only people who dependably give precedence to food quality over dogma are professional chefs. I learned of Kokekokko, my favorite Japanese yakitori restaurant, from a profile of David Myers in *Food and Wine*. Asked for his "current obsession," the chef-owner of Sona, an elegant Los Angeles restaurant, came back with this unassuming storefront in Los Angeles's Little Tokyo that I had driven past dozens of times without reason to notice, whose name means "cock-a-doodle-doo."

"For $25, you can get a chicken tasting menu," Myers told the magazine. Had it not been for that comment, I doubt I would have learned that one of the great restaurant meals in this city of great restaurants consists of marinated and skewered chicken pieces—breast meat, thighs, hearts, livers, wings, gizzards, meatballs, and more—cooked simply on an open grill; adorned with dabs of wasabi, mustard, or ginger puree; and served with the most intensely flavored chicken broth I have ever tasted.

David Kirby, a poet and English professor at Florida State University, goes in precisely the opposite direction for restaurant recommendations. Rather than search out the advice of chefs, he consults strangers on the street. "When you want to know where to eat, you should ask someone who looks like you, or a slightly better-fed version of yourself, maybe someone just a little paunchier than you are and a year or two older," he advises. By deploying this strategy in cities throughout Europe and North America, Kirby says, he has been led to some fabulous meals. And everyone he has asked for recommendations has happily complied. "Surely the fact that they're more or less looking into a mirror has something to do with the upbeat attitude, but I'm betting my new friends also recognize me as a frat brother of sorts, a fellow initiate in the freemasonry of food lovers," Kirby wrote in a *New York Times* article about his experiences.[2]

As M. F. K. Fisher wrote, "There is in some few men of every land a special hunger, one which will make them forgo the safe pleasures of their own beds and tables, one which initiates them into that most mysterious and ruthless sect, the adventurers." Whether through interviews with strangers, or what is more common, by way of friends and acquaintances, such food adventurers tap into "a pipeline of inside information," as Janelle Brown, a *New York Times* reporter, called it in her piece about people in their twenties and thirties who frequent ethnic eateries in shabby strip malls in Los Angeles.[3]

Reading Brown's story was a revelation to me—not about the dozen or so restaurants she mentioned, but about myself. Even though I am almost twice the age of the youngest person in Brown's story, apparently I am part of what she christened a "cult of underground cool." I'd already eaten nearly everywhere she listed, even Palms Thai, where an aging Elvis impersonator with a heavy Asian accent croons nightly and wild boar stew is a house specialty.

At another restaurant Brown featured in her story—Alegria, a Mexican restaurant a few blocks from our home that she singled out for its "inventive and delicate dishes"—my wife and I are longtime regulars.[4]

Another Kind of Hunger

Are those of us who seek out such places "true food fans," as Brown describes us, or are we lemmings who "eat where they're told and follow trends," as Jim Leff, founder of chowhound. com, the principal online forum for food adventurers, accuses? When I lunched with Thi Nguyen, an editor of the Los Angeles edition of chowhound's weekly newsletter, *Chownews* ("there is no better food tip sheet anywhere," boasts the Web site), he dismissed trendy ethnic restaurants in neighborhoods like mine out of hand. "The nearest outpost of non-white food near a very white area," Nguyen called them.[5]

True to chowhound.com's edict to "spurn trends and established opinions" by scouring neighborhoods "for hidden culinary treasures," Nguyen and his fellow food adventurers would sooner risk a bad meal at an unsung hole-in-the-wall than patronize a popular restaurant.

Much about food adventurers is undeniably endearing. In person and online, they almost never utter words like "carbs" or "cholesterol." Almost childlike in their enthusiasm for newly discovered restaurants and neighborhoods, they say things like, "You can look at the city from afar, you can theorize about it, but there's something about sitting down in a place, ponying up your five dollars, and just eating something really good. It makes me so goddamn happy" (Nguyen).[6]

This attitude does not necessarily result in better eating, though. In three years of checking out restaurants acclaimed on chowhound.com, the only first-rate meal I've had was at Lotus of Siam, a Thai restaurant in a run-down strip mall in Las Vegas that had been discovered not by chowhounds, but by the former restaurant critic for *Gourmet* magazine, Jonathan Gold. Back in 2000, Gold declared the place the best Thai restaurant in North America, and about a year or so later, Dave Feldman, a New Yorker who posts frequently on the Web site, organized a gathering of chowhounds at the restaurant.

Feldman scheduled the meal to coincide with a convention he was attending in Vegas, and I and a couple of San Francisco chowhounds flew in for the occasion. Sucuay and Saipin Chutima, the husband and wife who own Lotus, feted us on northern Thai cuisine that bears only scant resemblance to the pad thai–centered fare at most Bangkok-style eateries in the U.S. Lotus of Siam's version of chicken soup, tom-kah-kai, has twice the kick of my Jewish grandmother's East European recipe. Flavored with fresh chilis, tomatoes, greens, and herbs, it is blessedly untainted by the canned coconut milk that characterizes soups in other Thai restaurants. I also loved the kang hung-lay, a rich pork stew that is both sweet and spicy and served with sticky rice.

Few of the other ethnic places I have tried on the recommendation of chowhounds have merited a second visit. In most, the food was so oily or overcooked that it was nearly inedible. Which raises the question: Why would any sensible person rely on restaurant recommendations from anonymous enthusiasts online when they have connoisseurs like Jonathan Gold to guide them to outstanding restaurants such as Lotus of Siam?

The answer has to do with another kind of hunger, the yearning for community described in Robert Putnam's book *Bowling Alone*. The decline over the past few decades in traditional organizations like churches and synagogues, community organizations, and even bowling leagues has left many Americans yearning for alternative channels for hooking up with others who share their interests and attitudes. Chowhound.com serves that purpose for food adventurers. In my interviews with chowhounds and in postings on the Web site, tales abound of people developing close relationships. "I consider Kevin a great resource and a great friend. If he didn't make his post on Café Hiro, I wouldn't have met my fiancée," began one posting.[7]

But meeting or marrying fellow food adventurers is one thing; informed restaurant recommendations are another. If you are after the latter, the place to turn is not arbitrary postings on the Internet, but articles and books by the likes of Jonathan Gold, who currently writes for the *Los Angeles Weekly*, or Ed Levine, R. W. Apple, or Peter Meehan of the *New York Times*, Robert Sietsema of the *Village Voice*, or Dara Moskowitz of *City*

Pages in Minneapolis (just to name those I read regularly). These folks pound a lot of pavement, eat a lot of lousy food, make a serious study of the diverse cuisines they write about, and unlike their colleagues who review high-end establishments, almost never get recognized or catered to by restaurateurs.

Their reviews have led me to scores of outstanding eating places in my hometown and on the road, including some utterly unexpected finds. Thanks to Moskowitz, I ate in what is, so far as I know, the only Kurdish restaurant in the United States. Located in St. Paul, Minnesota, Babani's makes a delectable eggplant dish called "Sheik Babani," stuffed with spicy ground beef and vegetables in a red sauce. Babani's rendition of chicken soup, called "Dowjic," is also a treat. Highly seasoned but cooled with lemon and yogurt and thickened with rice, it is unlike any other I have tried.[8]

These reviewers are also fun to read. Jonathan Gold is a master of description. Of one of his favorite Chinese restaurants, he wrote, "The aftermath of a dinner at Hua's Garden is like a Francis Bacon painting splashed across the tabletop in shades of red—gory puddles of scarlet juice alive with Sichuan peppercorns, scraps of scallions, and frog bones stripped clean of their meat." In another piece, he described fish at a Vietnamese restaurant as having arrived "mouth agape like Aaron Brown deprived of a Teleprompter."

For Extra Credit, Define Delicious

Professional reviews tend to be superior not only as literature, but in the criteria they use to judge a meal.

For our initial lunch together, Thi Nguyen selected a Colombian restaurant that had been described as "glorious," "fabulous," and "wonderful" in postings on the Web site. It was one I had driven by many times, and I was hopeful that it was in fact a culinary treasure. The moment I entered, I could see it met most of the criteria that food adventurers look for. The menu board was in Spanish, as were conversations at almost all of the tables; the decor was dilapidated but comfortable; and the owner waited tables.

The only thing lacking was what chowhounds call "delicious-ness." Given that the very definition of a chowhound, according to the Web site, is someone who "hates to ingest anything unde-licious" and "sniffs out secret deliciousness," this incongruity seemed odd. What someone on the Web site called "the best fried pork I have ever had" was, to my palate, the opposite.

Swallowing a leathery bite from that dish, I asked Nguyen about this notion of deliciousness. "Deliciousness is kind of an empty category. 'You should seek the delicious' sounds to me like 'You should do what is good.' It's almost a redundancy," replied Nguyen, a Harvard graduate enrolled by day in the Ph.D. program in UCLA's Department of Philosophy. So what, then, I asked him, do his fellow chowhounds mean when they use the term?

"Just as the word 'good' means 'that which you should do,' delicious is the stuff you really like to eat," he explained. "I'll tell you this, I like restaurants that have good food, as opposed to all the other ways a restaurant could be good. Maybe that's the firmest thing I can say."

For a food adventurer, all of the elements other than food quality that go into earning a restaurant three or four stars from the likes of a *New York Times* reviewer or *The Mobil Travel Guide*—service, wine list, ambience, and so forth—are strictly tertiary, a position with which I am partly sympathetic. More of-ten than I care to count, I have handed my credit card to adoring and adorable waitstaff in exquisite settings where the hundred-dollar dinner was not nearly as pleasing as any meal I've had at Phillips, a barbecue joint in South L.A. that has no tables, but where ten bucks buys a platter of smoky ribs "as beefy as rib roasts beneath their coat of char, tasty even without the sauce," as Jonathan Gold described them.

To me, a dive with great food deserves as much respect as any upmarket restaurant. If a reasonable definition of culinary greatness is the ability to take simple ingredients and transform them into dishes that please and astonish, places like Phillips Barbecue arguably merit *greater* acclaim. Three- and four-star restaurants may be more comfortable and serve better booze, but you can't always count on experiencing true culinary greatness.[9]

Those who reject this view hold that, as William Grimes argued in an article in the *New York Times* a few years ago, "perfection at one culinary level does not compare with perfection at a higher level." The former *Times* critic allowed as how "the perfect three-minute pop song cannot grip the imagination and hold it the way a three-minute polonaise by Chopin can. Subtlety, finesse and refinement deserve a higher score. Art trumps craft."

If to you "refined" means "well-bred," it is true that the plasticware at Philips cannot compete with the sterling silver at Le Bernardin. But using another dictionary definition of refinement—"improvement or elaboration"—a great barbecue place clearly qualifies.

Occasioned by a Zagat survey that ranked modest eateries in outer boroughs higher on food quality than some Manhattan citadels, Grimes's article triggered a downpour of disapproving postings on chowhound.com and other Web sites geared to food adventurers. "I'm sure Grimes was the sort who was relieved when the Beatles (or George Martin, I suppose) added strings so that he could, finally, revel in their subtlety, finesse, and artistry, and not feel like such a geek for not liking them," someone wrote on chowhound.com.[10]

But revealingly, in their critiques of what they considered Grimes's elitism, neither Feldman nor his compatriots championed the target of Grimes's attack, the Zagat surveys. Derived by tabulating votes cast by everyone who chooses to fill out a form (about thirty thousand people for the New York City edition), Zagat is the most democratic of review outlets, yet chowhound.com's founder Jim Leff derides diners who go to Chinatown "clutching their Zagat."

People who value the opinions of online amateurs over those of professional reviewers may *sound* like the ultimate democrats, but they can be snobs in their own right. Part of the appeal of seeking out obscure eateries, a New York chowhound named Mark explained to me, "is that you're discovering something unknown, which is kind of thrilling. You feel like you're in a select group of people who are in on something that the general public doesn't know."[11]

Mark made that comment during a gathering of chowhounds that Dave Feldman had organized on my behalf during one of my visits to New York, and I had to restrain myself from replying that I could not imagine why anyone would *want* to be among the select few who know about the restaurant where we were dining that night. I have not made a study of Spanish tapas restaurants in New York, but contrary to postings on chowhound.com declaring this particular place the best in New York, I have to believe there is somewhere that serves less bland renditions and cooks them in oil of a more recent vintage.

Indeed, *Zagat Survey New York* lists several candidates, and Robert Sietsema had written about others. Those won't do, however, for ardent food adventurers. To quote another of my dinner mates that night, "If everybody knows about it, there's nothing there to discover anymore."[12]

Inauthentic Authenticity

I may have considered the food at that restaurant second-rate, but the place met one of the cardinal conditions that food adventurers demand. It felt authentic. Located in the basement of the Spanish Benevolent Society on Fourteenth Street near Eighth Avenue, La Nacional catered to the Spanish expatriate community of New York.

For a food adventurer, to call a restaurant authentic is the ultimate praise, the equivalent of "healthy" for a follower of the gospel of naught, and to declare something inauthentic is the ultimate slam. "By 'bad' food we mean non-authentic," a posting on chowhound.com noted, and Jim Leff argues that when a dish is less than delicious, lack of authenticity may well be the reason. But like deliciousness, authenticity turns out to be an elusive concept. The commonsense definition, that authentically prepared food "tastes like it is supposed to taste in the country where it is made," as one of the participants at the Lotus of Siam lunch, Cheryl Pochapin, put it to me, begs the question of how faithful a dish must be to qualify.[13]

So does the seemingly simpler demand that cooks use only ingredients that are native to the region where the dish originated. While almost a cliché among advocates of authentic cooking, if that dictate were taken seriously, Italian restaurants would have to remove from their menus any dish with tomato sauce, and many of the spiciest dishes at Thai and Chinese restaurants would be verboten. Prior to Columbus's voyages, there were no tomatoes in Italy or chilis in Asia.[14]

To classify a dish as authentic because the person preparing it uses the same ingredients and cooking techniques employed at another time or place is to ignore that dishes come about in the first place, and change over time, as new ingredients and cooking equipment become available. Even a work of fiction can change a cuisine. Over the past decade, restaurants in Mexico City and the U.S. that specialize in precolonial foods have added a "traditional" rose-petal mole to their menus that actually comes from the 1992 novel *Like Water for Chocolate*. The recipe feels old-world, but in reality, it is a creation of the book's author, Laura Esquivel.[15]

Restaurant goers get their notions of authenticity mostly from restaurateurs who promote what will suit their own needs. In a study of Thai restaurants in Dallas, sociologist Jennie Molz cataloged a host of messages the eateries use to create what she calls "staged authenticity." Menus boast of "original Thai country cuisine" and use phrases like "a favorite Thai noodle dish." Another common ploy, Molz notes, is a warning that the food may be too spicy for American palates and can be made milder on request.[16]

Diners seldom assess the authenticity of Thai restaurants from knowledge of the cuisine as it exists in Thailand or in the homes of Thai émigrés in the U.S., Molz found, or even from cookbooks. Rather, they base their assessments on comparisons with other Thai restaurants. If a place serves spicy food that includes ingredients Americans have come to expect, such as coconut milk, lemongrass, and basil, and has the typical sauces on the tables and Asian-theme artwork on the walls, it qualifies. Ethnic restaurants "are formulaic, both responding to and creating their diners' expectations of authenticity," Molz observes.

A pair of researchers from Laval University in Canada point out another reason customers and owners tend to view ethnic restaurants differently. "For owners, economic matters clearly came first; the desire to preserve and disseminate their own respective cultures through the medium of the restaurant barely crossed their minds," Laurier Turgeon and Madeleine Pastinelli report in a paper on a study they conducted of immigrant restaurateurs in Quebec City.

Tellingly, none of the restaurateurs in that study had actually been involved in the restaurant business in their home countries. "Many immigrants become restaurant owners," Turgeon and Pastinelli explain, "for the simple reason that the host society asks for 'authentic' ethnic cuisine prepared by authentic ethnic people."[17]

Were authenticity buffs to base their beliefs on what appears regularly on tables in homes rather than on hearsay from other food adventurers and stagings by self-interested restaurateurs, they would have to revise their views fairly radically. They would have to give up some of the popular dishes in ethnic restaurants, for a start, because in their countries of origin these are special-occasion foods eaten primarily at festivals and holidays, or in homes of the rich. In most places in the world, everyday cooking is exactly as the name implies—pedestrian—and in the world's urban centers, it tends to include elements that American food adventurers reject as Western corruption.

"Most Shanghainese learn to make a potato salad remarkably similar to Western potato salad," someone who signs himself Gary pointed out in a chowhound.com exchange about authenticity. Gary's Chinese wife brought a potato salad recipe with her to the U.S., he reports, "without having ever been to a Western restaurant. I guess you could say she makes an 'authentic' Shanghai potato salad."

To culinary purists, there is no dirtier word than "fusion"; they share Octavio Paz's view of the melting pot as "a social

ideal that, when applied to culinary art, produces abomina-
tions." But ethnic cooks in the U.S. belie that generalization by
integrating European foods into their repertoires, as well as ingredients and cooking styles from other immigrant groups besides their own. At a Korean-owned Chinese-style restaurant in
central Los Angeles whose clientele is primarily Koreans and
Latin Americans, for instance, for about twelve dollars, including tip and tax, I regularly feast on a massive bowl of spicy seafood soup with velvety, hand-pulled, Chinese-style noodles
whose translated name is "three sea chowmamein." The meal
also includes a plate of delicate steamed dumplings with meat
filling, along with assorted kimchis made of fermented cabbage
and turnips, and a bottle of Tsingtao beer.

Korean-Chinese establishments abound in Korean neighborhoods in Los Angeles, along with Korean-Japanese places where
sashimi is served with hot sauces and bean pastes in addition to
the Japanese-style soy sauce and wasabi. On the opposite coast,
in Brighton Beach in Brooklyn, you can get good sushi in—of
all places—Russian restaurants. Some serve it Japanese style
alongside conventional Russian dishes; others offer Russian
variations, like caviar-enhanced sushi.[18]

At Basta Pasta, a Japanese-owned Italian restaurant in Manhattan, the menu includes spaghetti with fish roe and Japanese
basil. New York City is home as well to Chinese-Jamaican and
Indian-Chinese restaurants.

Contrary to some food adventurers' notions of a fusion-free
past, this mixing and matching is nothing new. Immigrant and
native groups in North America have been combining foods
from one another's traditions for at least four centuries. Donna
Gabaccia, a history professor at the University of Pittsburgh,
has documented numerous examples, from the addition of corn
to the diet of English settlers in the seventeenth century and the

incorporation of European fruits and grains in Native American cooking of the eighteenth to the popularity of German lager beer among multiethnic crowds in nineteenth-century saloons, and the nearly universal acceptance by the late twentieth century of the bagel, a food eaten mainly by Eastern European Jews a century earlier.[19]

The Adventurer Wears Prada

Food adventuring itself has a long and contentious history. Cross-cultural diners have variously been celebrated as pacesetters and denounced as perverse.

Present-day chowhounds continue a tradition that dates back more than a century. "Today the 'slummers' eat, drink, and are merry in their experience with strange new dishes," a young New York intellectual wrote in 1893 of a nine-course meal he and his friends enjoyed at a Chinatown restaurant that, he suggested, conventional Americans would find "as uninviting as a pig-sty."[20]

A few years later, in the early part of the twentieth century, artists and intellectuals frequented eateries run by another immigrant group. At Italian restaurants, for fifty cents they were served a pint of red wine, an antipasto, a pasta dish, a main course of fish or meat, vegetables and salads, tortoni or spumoni for dessert, and fresh fruit, cheeses, and strong coffee or espresso.

What those early food adventurers' meals at Italian and Chinese restaurants resulted in, aside from heartburn, has been a subject of considerable dispute, however. In one view, interethnic eating "brought new tastes, new sounds, new scents, new form, new colors, but above all, new feelings" to the native-born Americans who ventured into them, as Maria Sermolino, the daughter of the manager of a restaurant in Greenwich Village during that period, wrote in a memoir. "Restaurants have encouraged, even in periods of social and political conservatism, the crossing of formidable ethnic and cultural barriers," historian Samantha Barbas of Chapman University argues. "In search of cheaper, quicker, and more interesting cuisine, Americans have often suspended traditional racial prejudices and opened themselves to a range of diverse culinary and cultural experiences."[21]

In this laudatory view, while food adventuring may be "a slim thread on which to build cross-cultural understanding," as Donna Gabaccia characterizes it, as she also notes, "given the depth of American fears about cultural diversity, it is better to have that thread than not." The outlook of food adventurers contrasts favorably with that of "people who say, sometimes with pride, 'oh I never leave the neighborhood, we never go above 14th Street, we never go to Queens,'" as Eric Asimov, the longtime author of the *New York Times*'s "$25 and Under" column, described many New Yorkers he told me he meets.[22]

But every generation of food adventurers has had its detractors, too. For their forays into Chinese and Italian restaurants, the slummers, bohemians, and other culinary adventurers of the late nineteenth century and the early twentieth century were accused of depravity and "morbid curiosity." According to their critics, rather than advancing American civilization, they debased themselves and those who would follow their lead. "One can easily imagine the effect of the sights witnessed on the girls of tender years, unsophisticated and practically ignorant of the world and its wicked ways," a letter to the *New York Times* in 1905 suggested of whites who dined among those the letter writer referred to as "heathen" Chinese.[23]

Critics of today's culinary adventurers, while no less disapproving, take the opposite position. Rather than looking down on immigrants and trying to protect white girls from them, they seek to protect immigrants from self-serving whites. The gentler form of this critique accuses food adventurers of insincerity. Chowhounds may claim they are simply in search of deliciousness, but in reality, they're after what the late French sociologist Pierre Bourdieu called "cultural capital." In food-adventurer circles, eating exotic foods is a mark of sophistication and a way to elevate one's social status.[24]

This advantage is ill-gained, though, in the view of some. According to David Bell, a cultural studies scholar in Britain, "the endless safari of the cosmopolitan, searching out the exotic and the authentic, is essentially a predatory practice: the pillaging of resources, the scouring of habitats, the uprooting and repackaging of the foreign, the novel, the dangerous."[25]

Samantha Kwan, a doctoral student in sociology at the University of Arizona, is more accusatory still. "The consumption of authentic ethnic food is an attempt to identify with the oppressed rather than the oppressor," Kwan contends. "It is an attempt to assert that despite living in the comforts of the developed world, despite dressing in pricey imports from the less developed world, despite driving to the authentic ethnic restaurant in a massive, gas-guzzling sports utility vehicle, that despite all this, one still empathizes with the less fortunate."[26]

In my experience, neither caricature of food adventurers holds water; as a group, they are neither trailblazing progressives, as some suggest, nor the white devils that Kwan and others envision. I have yet to meet anyone who believes that by eating ribs or kimchi he or she identifies with the oppressed; and some food adventurers, like Thi Nguyen, the son of Vietnamese immigrants, are from disadvantaged groups themselves.

Contemporary critics of food adventurers also fail to see that restaurant revenues have provided immigrant restaurateurs with what Sylvia Ferrero, an anthropologist, has described as "the bedrock upon which they improve their social position." By marketing their cuisines beyond their own ethnic group, substantial numbers of immigrants have been able to move into the American middle class.[27]

Food adventurers suffer from a perceptual disorder we've seen before. They view foods and the people associated with them in black-and-white terms—good or bad, authentic or inauthentic, savior or scoundrel.

Going Mainstream

Ethnic restaurants whose clientele extends beyond food adventurers have to contend with even more conflicting demands. Those that attract a broader audience have people like Michael Jacobson looking over their shoulders. The pricey four-star establishments I discussed in the previous chapter are not targets for attacks by the Center for Science in the Public Interest, even though they serve high-fat, mega-calorie dishes. "They're trivial," Jacobson told me. "Almost nobody eats there." But if you operate a moderately priced Mexican, Italian, or Chinese restaurant, watch out.[28]

Jacobson's war on such places began in earnest with a publicity stunt in 1994 that nearly all the major media found irresistible. "Mexican restaurant food makes Chinese and Italian look good" by comparison, the Center for Science in the Public Interest's press release declared, and at a press conference, Jacobson held up test tubes filled with fat from chilis rellenos and pronounced the dish as bad as fettuccine Alfredo and kung pao chicken. Chilis rellenos have as much fat as four McDonald's Quarter Pounders, he stated.

The day after that press conference, in an article in the *New York Times*, Marian Burros coolly dismissed rejoinders from Jacobson's critics in the restaurant industry. Their arguments, she wrote, "were not easy to make in the face of visual proof of high levels of fat in Mexican food." But in point of fact, as author Richard Rodriguez pointed out, Jacobson's study was little more than jingoism. "Is there a more enduring slur against Mexico than 'dirty?'" asked Rodriguez. "Americans," he suggested, "have a taste for the clean and pure. Mexicans are more easy with all that is messy in life."

People familiar with Latin American eating patterns took exception on their own grounds. Dishes at Mexican restaurants are usually shared, they noted, by whole tables of people. The typical customer, even one who orders the highest-fat item on the menu, is unlikely to consume "a whole day's worth of fat from a single entrée," as Jacobson claimed.[29]

"Are we angry? Of course, the kitchen is angry. It's a kind of insult, you know, that Chinese food is unhealthy," an immigrant from Hong Kong who helps manage a family-run restaurant in Boston told a researcher after one of the CSPI's slams on Chinese food hit the press. Customers began making impossible requests, the restaurant's staff reported. They recalled one woman who demanded her meal be prepared without oil, sugar, soy sauce, or MSG. "But I also want it to be tasty," she told the waiter.[30]

Studies of ethnic restaurants document numerous examples of restaurateurs omitting ingredients from their dishes to accommodate the gospel of naught. Fried eggs, for example, routinely appear atop rice dishes in parts of Asia, but rarely at Asian eateries in the U.S. Even at places like Lotus of Siam, which appeal to food adventurers but actually make most of their living off of more mainstream customers, proprietors find they must adjust their dishes. Sucuay Chutima told me he studies the behavior and eating patterns of his American customers and feeds them accordingly. "We know you don't like fat floating on your curry like they do in Thailand," he said by way of example. "Thais like it because of the color, but over here, you see that as a sign of a fatty dish."

Chutima makes a distinction between what he calls "traditional Thai cooking," some of which must be abandoned to accommodate American sensibilities, and what he refers to as "the integrity of the dish," which has to do with the quality of the ingredients. "We don't cut corners on the ingredients, we just modify the process of cooking," he explained. "Instead of using a whole bunch of lard or fat in the curry, we just stir-fry it lightly in soybean oil. Stuff like that." The dishes taste somewhat different than they would in a traditional preparation, Chutima said, but "sometimes you have to adapt to the local taste."

Where to Find Consistent Chinese Food

Mainstream American diners impose another demand as well on ethnic restaurants. A dish must taste the same every time someone orders it.

In haute European cuisine, this requirement is readily accommodated. Precise adherence to established recipes is part of the craft, and consistency is highly valued. As Mario Batali once put it to an interviewer, "If someone has a great dish and returns to have it again, and you don't serve it to him in exactly the same way, then you're a dick."[31]

But in other cultures, other values prevail. In China, restaurant dishes are "conceived and prepared individually, each matching mastery and imagination with fine, fresh ingredients," reports food writer and novelist Nicole Mones from twenty-five years of travel there. She contrasts that approach with Chinese restaurants in the U.S., where chefs generally rely on a small number of premixed sauces and cooking methods to deliver dishes that taste identical time after time.[32]

In many of the places from which Asian American immigrant restaurateurs come, menus are governed by the availability of ingredients on a particular day or week. Yet paradoxically, although "fresh" and "local" have become bywords for high-end chefs in the U.S., American diners grant *less* freedom to chefs at ethnic restaurants for responding to seasons and markets. To ensure that their pad Thai tastes the way customers expect, chefs have to buy commercially produced sauces, and when bean sprouts or scallions are unavailable locally, they have to import them. Ethnic restaurateurs are forced to rely, in other words, on the large wholesale suppliers and global markets that food adventurers and activists deplore.

At a restaurant like Lotus of Siam, where customers expect authentic, old-country cooking *and* the flavor profiles they associate with Thai food, that means buying from both big suppliers and, for specialty ingredients, mom-and-pop stores in ethnic neighborhoods.

During periods of heightened paranoia among their American customers, the conflicting demands on ethnic restaurants for consistency and authenticity are greater still. "They want it steamed, they want the sauce on the side, and they don't want the salt, but they still expect it to be exactly what they've become used to tasting," an owner of a Boston Chinese restaurant complained to a researcher after the CSPI campaign. For a modest family-run restaurant like hers, balancing all of these conflicting demands is a nearly impossible task.[33]

Although no self-respecting food adventurer would be caught dead in the kind of restaurant that *does* have the wherewithal to deliver consistent dishes time after time while being highly responsive to its customers' preferences and pressures from the food police, these places have been instrumental in enticing Americans to move beyond the narrow confines of mainstream American cuisines.

Ethnic fast-food restaurants do all of that. The chain I have come to know best, Panda Express, has introduced hundreds of thousands of people to stir-fried Chinese food at its nearly eight hundred outlets throughout the country. Many of those people would never venture into a mom-and-pop storefront, though in important regards, Panda Express is an outsize version of the same thing. Owned and managed by a husband and wife who emigrated from China in the late 1960s, Panda has immigrant chefs in charge of menu development, and Chinese Americans in key management positions.

Panda Express restaurants may be located primarily in mainstream shopping areas, but a few years ago, when the company built a 178,000-square-foot corporate headquarters with offices, warehouses, and research, development, and training facilities, it chose a site in an area of the San Gabriel Valley outside Los Angeles that has one of the largest Chinese populations in the U.S. Within a short drive are hundreds of independent Chinese restaurants, a number of which appear on reviewers' and food adventurers' lists of the nation's best and most authentic. No one with whom I spoke at Panda headquarters pretends to be duplicating what is served at those places, but after spending time with Andy Kao and Sun-Fu Huang, the lead chefs, and Sonya Wong and Anna Yee, who run the marketing department, as well as the company's president and CEO, Peggy and Andrew Cherng, I came to feel they are doing something more difficult.

Panda Express is taking the emblematic cooking methods and some of the defining flavors and ingredients of Chinese cooking, and finding ways to make these appeal to a broad audience. During my most recent lunch at the Panda Express a few miles from my house, African American and Latina mothers and grandmothers ate with their preschoolers, and white truck drivers and store clerks from nearby malls ate alone or in pairs. At the massive food court at the Mall of America in Minneapolis last time I was there, the long line in front of Panda Express was markedly more blond, but with a smattering of families from the local Hmong community.

Another place Panda Express has been popular is Michael Jacobson's Center for Science in the Public Interest. Panda entrées make up fully one-quarter of the dozen dishes available in food courts that the CSPI approves. Although it disapproves of nearly everything that other mall and airport food concessions have to offer, the CSPI deems the Panda entrées "delicious and pretty healthful."[34]

Unexpected Authenticity

I, too, find the offerings at Panda Express impressive in some regards, though "delicious" is not a word that would come to mind. That the activists at the CSPI would choose that word says more about their limitations as restaurant reviewers than about the merits of the food Panda sells. This is not "Gourmet Chinese food," as the company's logo asserts. But it is no mean feat that, for about the same price as a Burger King Whopper and fries, at Panda I get a perfectly good meal of steamed rice, chicken with mushrooms, and tofu with eggplant. Both entrées are prepared by a wok chef on the premises, and because the company's research shows, in Andy Kao's words, "our customers want healthy and fresh," none of the vegetables come from cans or freezers.

Prior to Panda Express there had never been a successful chain of Chinese restaurants in the U.S., despite several efforts, some of them well financed. The last time a group of Chinese eateries drew mass audiences was nearly a century ago. Unlike Panda, which offers its customers choices from about a dozen entrées, those places sold essentially one dish: chop suey. At least two hundred chop suey houses came into being on the East and West Coasts in the early 1900s, a period during which a substantial portion of the American population became, as some called themselves, "chop suey addicts."

By midcentury, chop suey was a staple in the American diet. Although the chop suey houses themselves had long since disappeared by that time, the dish could be found on menus at mainstream diners and cafeterias, and in the canned- and frozen-foods sections of grocery stores throughout the country. (La Choy and Chun King, the top-selling brands back then, are now units within ConAgra.)

Modern-day food adventurers consider chop suey the archetypal faux-ethnic dish, but like the food at Panda Express, it was an important vehicle for changing the American palate. Thanks in large measure to the chop suey craze, as historian Samantha Barbas notes, "flavors and ingredients once considered exotic—soy sauce, bean sprouts, water chestnuts, ginger, among others—became an accepted part of the mainstream middle-class diet."[35]

In truth, neither chop suey nor the items on Panda Express's menu are as inauthentic as they're made out to be. The distinguished food historian Alan Davidson went so far as to label the standard story about chop suey an example of what he calls "culinary mythology." In the usual telling, the dish came into existence when a Chinese cook took revenge on an obnoxious customer by combining a mishmash of leftovers and giving it a

Chinese name that translates as "odds and ends." In some versions of the tale the badly behaved diner is said to have been a drunken miner; in others he is a San Francisco politician or another colorful Bay Area figure of the late nineteenth century.

Actually, though, Davidson reported, chop suey hails from Toisan, a rural district south of Canton from which many of the Chinese immigrants of the 1800s emigrated.[36]

The myth does have a kernel of truth: in the traditional preparation of chop suey, as in the earliest American versions, leftover or odd-lot vegetables were used. But the same can be said of some dishes that are held in high regard by gourmets. Many of the stews, soups, and salads canonized in cookbooks and served at fancy restaurants undoubtedly had their origins in efforts by cooks to make use of food that otherwise would have gone to waste. Several times when I have asked well-known chefs about the process of developing new dishes for their restaurants, they have responded that they began by assessing which meat and vegetable parts were left over from other dishes on their menus.

The dishes on Panda Express's menu also come, directly or indirectly, from China. Andy Kao gets ideas for new dishes and revisions of old ones, he told me, by exploring the offerings at restaurants and culinary competitions in Asian American communities in the U.S., and from trips to China. "If a dish is popular there, I take a note, and when I come back, I try to use that knowledge in our dishes," he says, "or use those Chinese ingredients in our sauces and those spices to create the aroma and the taste our customers will like. From the popular dishes in China and in Chinese restaurants here we get ideas, even if we don't use exactly the same ingredients."

The main changes, Kao and his colleagues tell me, are in the levels of spiciness, and adherence to written recipes. "Skilled

Chinese cooks like to use their own creations, add something here and there, but we actually want our cooks to follow our recipes," Peggy Cherng said.

Adapting traditional ways to accommodate the clientele hardly makes Panda's food non-Chinese, however, as Kao pointedly reminded me. "You know, in China, too, the chefs try to create new dishes to meet their customers' needs," he said.

Beef Tacos and Orange-Flavored Chicken

Why did it take so long for Chinese food to gain a foothold in the American fast-food market? Why didn't a Panda Express materialize in the 1960s and 1970s, the heyday of the fast-food industry?

In the answer to those questions lies a surprising truth about the mechanisms by which popular tastes in fast food change in the current era. Rather than being steered by food adventurers, celebrity chefs, or big-buck advertising campaigns by the major chains, ethnic chains succeed largely by piggybacking on established trends.

Mainstream America's definition of a proper meal having long been meat and potatoes, it was perhaps inevitable that the translation of that meal into the vernacular of fast food— burgers and fries—would predominate in the early years. But by the time we collectively ate our billionth such meal, we hankered for variety, and so in the late 1960s, a decade after Ray Kroc opened his first McDonald's and a period in which he was adding a thousand new outlets a year, alternatives began to emerge.[37]

The first Pizza Hut opened in Wichita, Kansas, in 1958, but the company did not have a national presence until a decade later. The main Mexican food chain, Taco Bell, also dates back to 1958, when the founder, Glen Bell, having run successful taco stands in Southern California for several years, set up a central commissary to supply the hundred restaurants he intended to open. But not until 1964 did he actually sell his first franchise.

In the second half of the 1960s, however, Taco Bell expanded steadily, and then in the early 1970s, much to my delight, the company really took off. A college undergraduate at that time, I invested a few hundred dollars in Taco Bell stock from money I had saved from a summer job. The stock quadrupled in price within a year or so, and I sold my shares and bought my first car.

The idea to buy the stock came to me during a cross-country trip with friends. Tired of eating burgers, and not wanting to slow our progress by searching for a decent diner, midway through our journey we stopped at Taco Bell out of desperation. I can't remember what we ate or whether we liked it; I remember only that as we drove away, I thought to myself that it would be just a matter of time before the masses demanded greater variety in their fast food, and this company had found a surefire way to provide it. Taco Bell's interpretation of Mexican food took the core ingredients of what McDonald's served—ground beef, cheese, and tomato-based sauces—but instead of putting them in buns, Taco Bell stuffed them in taco shells. Pizza Hut had done much the same, only with pizza dough.[38]

Chinese cuisine, on the other hand, is not amenable to burgerization, because neither ground beef, cheese, tomato sauce, nor bread is among its component parts. Chinese fast food became viable only once Americans began to worship chicken. Born in 1988, Panda Express expanded to five hundred locations during the 1990s doing what almost every other growing chain of that time did. It created likable chicken dishes. At a time when upstarts like Boston Market amassed fortunes selling nothing but "healthy" alternatives to the Kentucky fried standard, and McDonald's, Burger King, and Taco Bell scrambled to add chicken sandwiches to their menus, the best-selling item at Panda was orange-flavored chicken. Kung pao chicken came in second.

Peggy and Andrew Cherng faced greater technical challenges, though, than their competitors, who merely substituted chicken for ground beef in some of their sandwiches or roasted whole birds and sold them in halves or quarters. Another part of the explanation for why a successful Chinese fast-food chain was slow in coming lies in the fact that, as Peggy Cherng, who has a doctorate in engineering, modestly puts it, "preparation of Chinese food is a little more complicated." In a nation where few children learn wok skills at home or in school, finding the hundred new cooks Panda hires every month or so requires some doing. The training of a Panda wok chef takes eight weeks, and the company teaches its staff other skills as well that traditional fast-food chains would never entrust to store-level workers; vegetable slicing and sauce preparation, for example, are done at each restaurant.

Yet Panda manages to keep its menu prices competitive with traditional fast-food chains, which are evolving in some startling ways.

6

Restaurant Hell

The Dissing of McDonald's

A quarter century into a distinguished career that had won him a slew of Best Chef, Chef of the Year, and Humanitarian of the Year awards, Rick Bayless suddenly found himself roundly condemned by journalists and fellow chefs. The *San Francisco Examiner*'s food critic, Patricia Unterman, called him a sellout. *Newsweek*'s Gersh Kuntzman labeled him a hypocrite. On Egullet.com, a popular discussion site for chefs and foodies, Anthony Bourdain called Bayless a "pimp for the Evil Empire."[1]

To attract that degree of scorn, a celebrated chef has to do something his colleagues and food writers find truly repugnant. Starring in disreputable TV shows, dressing bizarrely, even confessing to out-and-out debauchery in the kitchen—none of those suffice, as Rocco DiSpirito, Mario Batali, and Bourdain himself have amply demonstrated.

The one line a chef must not cross is into fast food.

Beloved by food adventurers and professional critics alike for his public-television program, *Cooking Mexican,* his Chicago restaurants, and several award-winning cookbooks, Rick Bayless had reason to expect a more respectful reception for his appearance in late 2003 in advertisements for Burger King's "Santa Fe Chicken Baguette Sandwich." Having brought new sophistication and respect to the food eaten by Mexican villagers, he might have thought that his efforts to do something similar for mass-market American food would be welcomed as well.

Admittedly, to some extent Bayless set himself up for ridicule. A founder of Chefs Collaborative, whose mission statement calls for member chefs to promote "sustainable cuisine by celebrating the joys of local, seasonal, and artisanal cooking," Bayless had portrayed himself as the antithesis of a fast-food eater. Just weeks before his Burger King ads ran, in a newsletter he sends to customers and journalists, Bayless published a piece titled "Rick's Secrets to Good Food and Healthy Living." In it he proclaimed that the "cornerstone of my everyday diet" is "fresh, not processed food."[2]

In his own defense, Bayless subsequently described himself as "both eco-chef and fast-food supporter." In a letter posted on his Web site, he argued that by appearing in the Burger King ad, he "encouraged a few hundred thousand people to experience a tasty, less-processed sandwich."[3]

And sure enough, the Santa Fe sandwich I was served at a Burger King near my neighborhood was, as Phil Vettel, the veteran restaurant critic at the *Chicago Tribune* admitted in a piece about the Bayless flap, "pretty good." A freshly baked baguette stuffed with roasted red and green peppers and onions atop a moist chicken breast fillet dressed in a spicy tomato sauce, the sandwich seemed to me a bargain at $2.99.[4]

Beware Whom You Call Stupid

I come neither to praise fast food nor to bury it, only to question its easy portrayal as the root of all evil. If, as we've seen, glorified restaurants are not always as glorious as their idolaters would like to believe, neither are demonized places as demonic as their vilifiers contend.

Nor are the people who work for them. Contrary to the impression given in advertisements for the Santa Fe sandwich, it was created not by Rick Bayless, but by someone at the fast-food company's headquarters in Miami. "Burger King brought the sandwich to me, and they asked me what I thought of it," Bayless recalled. "I was pretty amazed at what they were able to accomplish, and I said, 'Well, I applaud it.' And they said, 'Would you endorse it?' "[5]

The man who actually developed the sandwich, while not known to the general public, is a highly accomplished cook who is as serious about food as almost any high-end chef. Valedictorian of his class at the Culinary Institute of America, Peter Gibbons worked in several areas of the food industry before becoming director of research and development at Burger King. In his off hours, he consults his collection of more than one thousand cookbooks and prepares dinner for his family. "My wife and I have an agreement. When we got married, I promised never to touch laundry and she promised never to touch food," Gibbons told me. "When I'm going out of town I leave dinners prepared so all she's got to do is finish them or put them in the microwave."

Gibbons considers the food he creates at Burger King as tasty and respectable as what he makes at home or, for that matter, what he is served at his favorite local Italian restaurant. Whether creating dishes for one's wife and child, for neighborhood regulars, or for the masses, he says, the measure of success is the same: "the best quality that's conceivable for that environment." In each venue, you approach that level of quality by understanding the likes and dislikes of the audience, the technical aspects of the food preparation, where to find the right ingredients, and optimal use of the equipment and personnel available in the kitchen.

Nowhere is all of that harder, Gibbons argues, than in a large fast-food franchise. "It's really, really easy to come up with something that's pretty esoteric, where there are no benchmarks, no one knows whether it's good or bad because they have no point of difference to compare from. What's really tough, and the personal challenge that I see in the job that I've got here," he told me, "is that everybody knows what his favorite hamburger is all about. For me to be able to say, I've made a new hamburger or chicken product that has sold in the tens of millions, that I've made tens of millions of happy customers, that's something that not a lot of people can say."

When we dis fast food, we dis everyone associated with it: Gibbons and his colleagues who develop it, the people who prepare and serve it, and those who eat it, few of whom deserve the snub.

Jennifer Talwar, a sociology professor at Penn State, spent four years working at Burger Kings and McDonald's around New York City and interviewing more than one hundred of her co-workers, most of them immigrants from China and Latin America. Time after time she heard stories of customers, employers, and neighbors—even strangers on the street—ridiculing and harassing those who work in fast food. Many of the employees she met went to considerable lengths to avoid the stigma. They lied about where they worked, took jobs in restaurants far from their own neighborhoods, and put on their uniforms only once they arrived at work, rather than wear them in public.[6]

Much as knowledge of obscure ethnic restaurants affords an aura of cultural superiority to food adventurers, putting down fast-food workers can do the same for folks whose own career prospects are limited—and what is particularly unseemly, for self-styled activists. In an article originally published in 1994 in the hacker magazine *Phrack* and quoted and reprinted around the Internet ever since, someone calling himself Charlie X recommended a series of tactics his readers could use to "screw over" their local McDonald's. Calling it "a given" that fast-food employees were "stupid enough" to fall for the pranks, he advised putting hair in a burger and getting the kitchen staff in trouble by complaining to the manager; slowing down the cashiers by having them repeat the contents of orders several times; and shouting obscenities at the drive-through window "to piss the employees off."[7]

Activists bad-mouthed customers as well. My own observations confirm what Elspeth Probyn of the University of Sydney refers to as "activists' infantilizing of the average consumer of McDonald's as someone who obviously cannot think for him/herself and has no control over his/her appetite and actions." In reading and listening to critics of fast-food companies—online, in print, on-screen, and one-on-one—I am frequently struck by the contempt they express for those who eat fast food, people they view as dopes.[8]

Customers themselves tell a different story. They give sound reasons for where they choose to eat, and they appreciate what fast-food restaurants have to offer. "McDonald's means everything to my son and me," a woman told a university researcher who interviewed people as they waited in line at McDonald's restaurants. "My husband left us a few years ago and it's just me and my son now. I work different shifts over at the hospital and don't have much time to cook."[9]

The benefits to parents are hard to deny. Taking children to most restaurants "is not something to be attempted without Prozac, a mobile baby-changing unit, and good insurance," Frank McNally, a writer for the *Irish Times*, noted in a column. "By contrast, a visit to a fast food restaurant is almost stress-free, thanks to a range of parent support services including wipeable surfaces, unbreakable food containers and toys specially designed to stop your children fighting with each other for several minutes at a time."[10]

Concurring with a reader who had told him that parenthood gave her a new appreciation for fast food, McNally posited, "If it's a choice between giving your family a healthy, balanced diet, and grabbing a few minutes peace for yourself, any normal parent will opt for the few minutes peace—reasoning that parental sanity is also important to the children."

The Ultimate Populist Place

Fast-food restaurants oblige children, too. At these places, unlike everywhere else they eat, children have the ability to order their own food and permission to eat it with their fingers, and they get to play while they eat.[11]

Constantin Boym, a Russian émigré who owns a design studio in New York City and teaches at the Parsons School of Design, holds that immigrants favor the big fast-food chains for similar reasons to children's. Recalling his own delight in discovering McDonald's soon after he moved from Moscow to Boston in the early 1980s, Boym observes that "immigrants, like children, are conscious of making the wrong gesture, looking funny or different, standing out in any conspicuous way."

The steps involved in eating in most restaurants—from asking for a table to conversing with a waiter and figuring out a tip—while taken for granted by locals, actually require a great deal of cultural knowledge and language skills. "In this respect," says Boym, "McDonald's is the ultimate populist place."[12]

Only at fast-food places do the poor and the middle class eat the same food under the same roof. Even the most distrusted group in society—adolescents—is welcome here. In recent years, teenagers in seemingly unlikely places like China and France have flocked to American fast-food chains for some of the same reasons as generations of Americans. In China, where teenagers have few opportunities to hang out unsupervised, McDonald's has become a popular locale with high school students during their off hours. For French teens, who have many more options for places to congregate and eat, American fast-food places have held a different sort of appeal. They break with tradition. When Rick Fantasia, a sociologist from Smith College, interviewed French adolescents about their fondness for fast-food restaurants, they told him they like the self-service aspect and the absence of table settings, utensils, and the usual rules of etiquette. You can talk loudly and make a mess, they said.[13]

In *Super Size Me*, his anti–fast food film released in 2004, Morgan Spurlock eats three large McDonald's meals a day for thirty days and vividly demonstrates that such a regimen is enough to make a healthy thin man fat and enfeebled. But from the vantage point of some less fortunate folks, the picture looks quite different. Having spent most of his savings and unable to find a job, Les Gapay, a former *Wall Street Journal* reporter in his mid-fifties, gave up his apartment and moved into the only shelter he had left. "One of the most difficult aspects of living out of my truck," he reported fifteen months into his ordeal with homelessness, "was finding places to go to the bathroom or just to sit during part of the day. I quickly learned the ropes. I often ate at fast-food joints because of the $1 promotional items. Two of those made a meal."[14]

For Gapay and thousands of other homeless people, fast-food places are safe places in which to warm up, while away the hours, and get a hot meal. When I hear activists and food snobs bemoaning the frequency with which low-income Americans patronize fast-food chains, a famously sardonic observation made by Anatole France in the late nineteenth century comes to mind: "The law, in its majestic equality, forbids the rich as well as the poor to sleep under bridges, to beg in the streets, and to steal bread."

No one disagrees that the poor, like most of the rest of the population, would do well to eat more fruit and veggies, but where else, for a few bucks, can a person of modest means get the complete, tripartite American meal (meat, potatoes, and vegetable), in a clean setting, with toys and diversions for the kids thrown in at no extra charge? Or should low-income Americans be forced to subsist on the Department of Agriculture's "Thrifty Food Plan," whose recipes, even if followed slavishly, are barely lower in fat and additives than a Quarter Pounder dinner with small fries and a salad, but require hours of shopping and preparation and don't taste nearly as good?[15]

I wonder also if middle-class parents really protect their children by encouraging them to imagine the food at McDonald's as akin to crack cocaine. In experiments at Penn State University, youngsters were fed large lunches and then offered junk food. Some ate a great deal of the junk food even though they were already full, while others ate almost none. What predicted how much junk food they consumed? Whether their parents forbid high-fat, high-sugar foods in their regular diet. These studies and others find that when children are told that a food is bad for them, they assume it must taste good, and they develop an appetite for it.[16]

Making Burgers Safe for the Suburbs

Well-meaning reformers have long attempted to revamp the dietary preferences of children and commoners. A century ago in England, socialists found it difficult to imagine why low-income mothers did not respond to repeated lectures about the nutritional benefits of serving porridge to their families for breakfast. That the dish required constant attention to prepare correctly and came out burned and foul-tasting when made by someone who had to attend to multiple tasks like caring for several children apparently never occurred to them.[17]

In 1917, Bailey Burritt, the director of America's largest relief agency at the time, declared that the cause of malnutrition was ignorance, not poverty. Pamphlets from his organization, the New York Association for Improving the Condition of the Poor, instructed slum dwellers that "overeating is as harmful as undereating" and urged them to stop spending so much money on meat. "Buy the right things with the money you spend," the association directed.[18]

Beef, in particular, has often been high on reformers' lists of the wrong foods for people of limited means to buy. "To most workers, eating better food, usually more meat and particularly more beefsteak, was one of the major rewards of hard work and a respectable job. To the reformers, this was simply a source of frustration, a product of the improvidence and ignorance of the working classes," historian Harvey Levenstein reports of an influential group of American nutrition activists in the late 1800s.[19]

Do-gooders have inveighed against different cuts of beef for a variety of reasons. Until recently, they urged poor and working folk to forgo steak on account of its cost, not because they considered it unhealthful. On the contrary, from the mid-nineteenth century to the late twentieth, the pricier cuts of beef were generally considered "very satisfying to the stomach and possessing great strengthening powers," as M. Tarbox Colbrath, the author of an 1882 cookbook, put it. Regularly consumed by much of the middle and upper classes at supper, "beefsteak deserves the highest rank among breakfast fares" as well, Colbrath proclaimed.[20]

Inexpensive cuts of beef, on the other hand, had a bad reputation during this period. Americans harbored what Edgar Ingram, founder of White Castle, the nation's oldest fast-food chain, called "a deep-seated prejudice against chopped beef." Viewed as a foul admixture of substandard meat and chemical preservatives such as sodium sulfite, and sold at county fairs, at lunch carts outside factories, and from run-down shacks, hamburgers were considered unhealthy. It was Ingram's genius to clean up their image by choosing a name that connoted purity and strength, and an architectural design for his eateries modeled on Chicago's Old Water Tower. The restaurants were covered inside and out with sparkling porcelain enamel panels to make them appear, in Ingram's words, "white with purity."[21]

White Castle's highly successful advertising and public relations campaigns throughout the 1920s boasted of meat delivered fresh two to four times daily from butchers who worked only with high-quality beef "cut and recut so that the food cells in the meat are not crushed." The chain was endorsed by a food scientist who vowed that "a normal healthy child could eat nothing but our hamburger and water, and fully develop all its physical and mental faculties."[22]

To further reassure the public, Ingram made known that all of his employees were young men of good character who had passed health exams, and he dressed them in spotless white shirts, white pants, white aprons, and white linen caps—all washed and pressed by the company at no charge to employees.

By 1930, White Castle had truly become, as Ingram bragged, "a national institution," with fast-food hamburgers available at the company's outlets and those of its imitators in most cities throughout the U.S. Frequented principally by working-class men during the 1920s, in the 1930s, these places successfully attracted middle-class men and women as well. The true heyday for fast-food burgers would not come, however, until the 1950s and 1960s, when, as Harvey Levenstein noted, McDonald's founder Ray Kroc "tapped into something that had fueled the rise of the earlier chains, American concern for restaurant hygiene," this time to entice suburban couples and their baby boomer children to McDonald's.[23]

At least as obsessed with cleanliness as Ingram, Kroc liked to recount stories of watching his grandmother scrub her kitchen floor, "which was already as clean as a nun's cowl." In his 1977 autobiography, *Grinding It Out: The Making of McDonald's,* Kroc ran photos of himself hosing down the walkway outside a McDonald's and boasted of hiring "fussy and fastidious" managers for his restaurants. "If I had a brick for every time I've repeated the phrase *QSC and V*—Quality, Service, Cleanliness, and Value," Kroc wrote, "I think I'd probably be able to bridge the Atlantic Ocean with them."[24]

Reasonable people can disagree about the quality of the food McDonald's serves. The preeminent Spanish chef, Ferran Adrià, was being honest when he said to a *New York Times* reporter about McDonald's food, "Ferran Adrià and the 100 best chefs in the world cannot do better for the price." But gourmet fare it's not.

And having waited in long lines at some locations, I can attest that the service sometimes falls short. Critics go off the deep end, though, when they fault McDonald's and its rivals on cleanliness and value, the last two items on Kroc's list. In his book *Fast Food Nation*, Eric Schlosser pathologizes the high level of hygiene at McDonald's as an outgrowth of Kroc's "obsession with cleanliness and control," even though, by Schlosser's own account, "the enormous buying power of the fast food giants has given them access to some of the cleanest ground beef."[25]

Schlosser's investigative reporting has helped prod fast-food companies to ensure more humane treatment of animals and better conditions for slaughterhouse workers and migrant farmworkers by their suppliers. But when he depicts the eating of fast food as "a form of high-risk behavior" on account of the threat of food poisoning, he misses the mark. Outbreaks of food poisoning in these places have been incredibly rare. "When I hear people say the meat is bad, I laugh. I've been in the business for over twenty-five years and I don't know of a single example of tainted food," the owner of several dozen McDonald's told me. (In a legal battle with the McDonald's corporation over a financial matter at the time of our interview, he was willing to speak to me only if I promised not to identify him.)[26]

This franchisee's claim that McDonald's runs "the cleanest restaurants in the world" is barely hyperbole. The long lists of safety regulations the company imposes on its suppliers and operators, and frequent visits by company inspectors—scheduled and unscheduled—guarantee as much. As Warren Belasco, a prominent food scholar at the University of Maryland, points out, when critics of fast food, waxing nostalgic about the disappearance of small-time restaurateurs, assume that those operators did a better job of safeguarding their customers' health, they neglect a great deal of evidence to the contrary.

Reports of meat patties made of rodent parts at modern-day fast-food chains are urban legends, unlike George Orwell's accounts in *Down and Out in Paris and London* of the "year-old filth in the dark corners," cockroaches in the bread bins, chefs spitting in soups, and Orwell's kitchen colleagues laughing at him for washing his hands before touching food. "Roughly speaking, the more one pays for food, the more sweat and spittle one is obliged to eat with it," Orwell wrote in 1933, foreshadowing reports by present-day chefs, researchers, and writers who have worked or observed in kitchens in independently owned restaurants where, as Orwell put it, employees "had no orders to be genuinely clean, and in any case no time for it."[27]

It's All McDonald's Fault

Amazingly, critics denounce fast-food chains for the last of Kroc's four guiding principles as well. Low prices are a bad thing, they say in all seriousness, both for the companies and for their customers.

In 2002, when McDonald's stock plunged to its lowest level in nearly a decade, the *New York Times* asked management consultants and restaurateurs what advice they had for the company. Wolfgang Puck's response: charge a dime more per burger. "You just can't make a good burger for 99 cents," he said, as if no one had enjoyed any of the billion-plus burgers the company had sold for that price or less and a 10 percent price hike would be no burden for McDonald's many low-income customers.[28]

By purveying cheap food, fast-food companies actually do a disservice to poor and working people—and the world at large—critics contend. The food isn't *really* cheap, their argument goes, once you factor in government subsidies to fast-food companies for hiring disadvantaged workers, and the health and environmental costs of industrially produced, illness-inducing ninety-nine-cent burgers.

But even if one accepts this diagnosis, why assume, as these critics do, that an appropriate remedy is higher prices? Wouldn't a more effective and humane option be to shift government subsidies from offending sectors of the food industry to those that produce the sorts of food the critics champion? If the point is to get fresher, leaner, less processed, more environmentally friendly food in the mouths of folks who now go for Whoppers, surely the way to do so is by making it more affordable and widely available.

"It's a measure of how astonishingly far we have come from the hand-to-mouth existence of our forebears," *New York Times* columnist Daniel Akst comments, "that rock-bottom food prices, once a utopian prospect, are now seen as a threat to the well-being not just of Americans but of countless unwitting foreigners who don't know enough to temper their relief at not having to go to bed hungry."[29]

The condemnation of Value Meals by people who don't have to worry how they will pay for their next supper is a measure, too, of critics' inclination to portray anything that McDonald's et al. do as satanic, regardless of how well received by the very people the critics say they want to protect. More than a few teenagers have been able to buy cars on the money they earn at fast-food chains—a bad thing according to Eric Schlosser, who complains in *Fast Food Nation* that "as more and more kids work to get their own wheels, fewer participate in after-school sports and activities." For younger children, McDonald's offers seesaws, slides, and climbing gyms at eight thousand of its restaurants. Failing to acknowledge how valuable these are for tens of thousands of families who don't have access to safe and well-equipped playgrounds, Schlosser sees them only in a marketing context, as lures to incite children to beg their parents to bring them in.[30]

Critics blame fast food for every modern-day ill from heart disease and cancer to crime and urban sprawl. According to an article in the *Ecologist,* "no corner of the Earth is safe from its presence and no aspect of life is unaffected." Joe Kincheloe, a professor of education at Brooklyn College, actually suggests that McDonald's is responsible for Japanese children's losing their facility with chopsticks.[31]

McDonald's engages in a "cultural pedagogy" that produces "disciplined subjects" with "colonized desires" and a "commodified identity," Kincheloe explains in a book published by Temple University Press that is part of a large body of writing by academics who use rarefied language from social and literary theory to discuss "the McDonaldization of society," as the title of a much discussed book put it. To read these books and articles is to be transported into a world where the McDonald's Corporation dominates the modes of thought and operating procedures of every aspect of human life.[32]

The word "hegemonic" appears seven times in the space of two pages in Kincheloe's tome, but real people and their actual points of view are nowhere to be found in much of the discussion. He dismisses as naively self-deluding the fast-food customers he interviewed who told him he was making way too much of what for them is a cheap, fulfilling meal. "People like you are scared of hamburgers," one young man said. "McDonald's has power because it makes money selling people what they want."

Legitimate questions can be raised about how McDonald's uses that power, of course. As the largest buyer of beef and the largest employer of minimum-wage workers in the U.S., McDonald's has inordinate influence over the cattle industry and many of the nation's most vulnerable workers, so it is entirely reasonable that activists and journalists direct attention to the company's treatment of animals and low-wage workers. But why the fixation on fast food over other industries, and the McDonald's Corporation in particular? In any sector of the economy, the largest firms are going to have extraordinary power over their suppliers and the sectors of the labor market they employ. If the issue is labor practices, why not focus on large firms that arguably behave worse than McDonald's: those that avoid paying even the minimum wage by outsourcing work to poor countries, for example, or those that employ fewer Americans of color, fewer women, and fewer disabled people?[33]

If environmental sustainability is a concern, why not begin with companies that deplete the planet's resources and pollute the air and water in order to produce products that decorate people's bodies rather than feed them? Jewelry, for example. McDonald's foes like to repeat that the company produces over a million tons of packaging annually. Yet according to Earthworks, an environmental group, twenty tons of waste are generated in producing a single gold ring. A gold mine in Papua New Guinea called Ok Tedi generates two hundred thousand tons of waste per *day*, Earthworks reports.[34]

The preoccupation of intellectuals and activists with McDonald's is partly a legacy of the political struggles of the 1960s and 1970s. In recent years, Joan Kroc, the widow of McDonald's founder, has given hundreds of millions of dollars to causes dear to progressives, like National Public Radio and a peace studies program at the University of Notre Dame. But her late husband had a different attitude. An avid supporter of the Vietnam War, Ray Kroc complained in his autobiography of universities populated with "phony intellectuals" who taught young people "a lot about liberal arts and little about earning a living."

Kroc refused to give money to colleges, because, he said, in America "there are too many baccalaureates and too few butchers."[35]

Activists who dared to criticize McDonald's Kroc dismissed as hysterical fanatics whose true target was capitalism itself. But the reality, then and now, is the opposite. Were capitalism the target, any Fortune 500 corporation would make an excellent choice. Not even in the top hundred companies on that list, McDonald's is far from the largest or the most powerful force in global capitalism. What sets it apart and makes it a magnet for dissent, aside from its reactionary founder, is the company's incorporation of capitalist values in the very products it sells. Management at most major firms espouses the virtues of speed, efficiency, consistency, and bigness, but at McDonald's, not only are those capitalist values drummed into employees; customers ingest them with every Big Mac. Served identically and relatively swiftly in Brooklyn and Beijing, and possessed of six hundred calories *without* fries, drink, or dessert, the Big Mac is the capitalist ethos in a bun.[36]

A Paean to the Cheeseburger

A card-carrying member of the Slow Food organization, I'm all in favor of meals that embody an alternative set of values. Given the choice of a big, hurried, utterly predictable meal at McDonald's, or a leisurely supper of small, fanciful dishes at a neighborhood restaurant, I'll almost always opt for the latter. If the restaurateur buys from small farms and supports the protection of endangered plants and animals, as Slow Food further encourages, all the better.

Making those kinds of meals available and appealing not only to well-to-do gastronomes but to the general population is a goal I support. But I also support another of Slow Food's guiding principles, one that has an ironic implication when applied to McDonald's cuisine. We members are called upon to "recognize food as a language that expresses cultural diversity" and to help "preserve the myriad traditions of the table."[37]

Unmistakably an American tradition and an expression of our culture, fast-food hamburgers merit a degree of deference. In the age of Panda Express and Baja Fresh, when even at McDonald's and Burger King the menu boards are packed with salads, chicken dishes, and other nonburger items, the Big Mac sometimes seems to me as old-fashioned as dishes at the Texas Chili Parlor in Austin, Texas, or any of the other retro restaurants on Slow Food's "where to eat" lists.

"The priorities of fast food already seem as outmoded as Futurism or Vorticism: they belong to an already bygone age. The fifteen-second hamburger will join the fifteen-cent hamburger: consigned to the dustbin of history," historian Felipe Fernandez-Armesto contends. If he's proved right, Americans will have lost something akin to the Golden Delicious apple: the most commonly consumed if least sublime variety of a remarkable class of food. As the food historian and cookbook author Elisabeth Rozin has argued, the common cheeseburger "is primal in its capacity to evoke a collective—and positive—human experience."[38]

In an essay eulogizing the cheeseburger, Rozin observes that the centermost component, red meat, possessing more of the nutrients humans require than almost any other food, has been almost universally prized by humankind. Only in recent centuries and select regions of the globe, however, has fresh red meat been widely available to anyone other than elites. Beef in its chopped form gained popular acceptance more recently still, as we have seen, but it provides, Rozin notes, "a genuine fulfillment of that atavistic craving in all of us for tender roasted meat running with fat and juice, a hunger that seems to have been a common part of our shared experience as human beings."

Rozin examines more parts of the cheeseburger than anyone would have imagined possible. Even the housing enchants her. "Parted slightly to reveal some, but not all, of its juicy cargo, offering a tantalizing glimpse of anticipated delights and hidden surprises," the bun satisfies primal longings in its own right as well. One of our species' earliest and most satisfying technical achievements, bread has played an important role in the human diet for centuries. The modern burger bun, while not the most flavorful or nutritious of breads, is no exception. Artisan bakers and carb-phobes may view this ultraprocessed product with horror, but for the rest of us, "its golden brown shape, round and puffed, is a promise of homespun richness, its lack of corners and hard edges an indication of ampleness and generosity, of unconstrained fullness."[39] Not only that: the bun gives us the chance to get up close and personal with the hot meat and cheese and gooey condiments without spattering our clothes.[40]

Those of us who haunt specialty cheese shops in search of France's finest may snicker at the low-quality American cheese in fast-food burgers, but Rozin argues that the slice is there for a good symbolic reason. It denotes plenitude. Viewed through a historical or cross-cultural lens, eating cheese and meat together is an incredible indulgence. Some religious and cultural traditions explicitly prohibit mixing the two, and as Rozin points out, in cultures that permit it, few people have had the means to dine on both of these nutritionally dense foods in the same meal, never mind in the same dish.[41]

Then there's the burger's chief condiment. At once sweet, tangy, salty, and spicy, ketchup offers flavors for every taste. Like the U.S. population itself, ketchup brings together elements and values from around the world. It is named after a nineteenth-century Chinese condiment (ketsiap); two of its principal ingredients—tomato and vinegar—come respectively from South America and Western Europe; and in deference to taste preferences of the English, sugar is added to cut the intensity of their strong flavors.[42]

With so much going for it, no wonder the hamburger was America's national dish for the better part of a century, beloved by people of every age, race, class, gender, and region. Corporate executives and line workers, children in school lunchrooms, retirees in rest homes—throughout much of the twentieth century it was difficult to find an American who didn't dine on hamburgers at least a couple of times a week.[43]

By the late 1970s, the golden age of the burger had begun to wane under numerous pressures. The food police demonized it, multiculturalists challenged the very notion of a national dish, and pizza, chicken, and Mexican chains marketed inexpensive alternatives. We still consume lots of hamburgers, but lower-income Americans eat far more ground beef than richer Americans, who may enjoy the thirty-dollar burger at Daniel Boulud's DB Bistro Moderne but who, like their predecessors in the late 1800s and early 1900s, discredit plebeian burgers as the opiate of poor and poorly disciplined people.

In and Out of Work

There is, however, a fast-food burger chain that has been spared the wrath of the food snobs and anti–fast food activists. The food at In-N-Out, a two-hundred-unit West Coast operation, is regarded as safe treyf, if not positively sacramental, even though it is every bit as fat-filled and fattening as what McDonald's sells. So cool is In-N-Out's food that *Vanity Fair* serves it to movie stars at the magazine's post-Oscars party at Morton's.

Where the McDonald's Corporation "symbolizes the homogenization of America and the dark side of globalization," In-N-Out represents "decidedly simple values," a writer for *Los Angeles Magazine* reports. "They're great," Eric Schlosser has declared. "Food with integrity," he hailed the four-item menu of fries, hamburger, cheeseburger, and "Double-Double" burger, the last of which supersizes the Big Mac by seventy extra calories and eight grams of fat.[44]

Fast-food critics' criteria for what qualifies as safe treyf can appear arbitrary and inconsistent to those of us outside the faith. In-N-Out gets the nod for treating its restaurant workers better and serving food made from scratch. But doesn't the latter contradict the former? Call me cockeyed, but I fail to see how In-N-Out's employees are advantaged by having to pull the lettuce apart and peel and dice the potatoes by hand rather than work with machine-cut ingredients. Nor do I find that those labors pay off in significantly better-tasting food. In-N-Out's beef patties have more fresh meat flavor than McDonald's, which is commendable; but only on the Double-Double—or, God forbid, the "4 by 4," a popular off-menu option with four patties and four slices of cheese—is there enough meat for the difference to be perceptible amid all the fixins. And though the words "Fresh Potatoes" appear in large letters, twice, on the paper tray in which In-N-Out serves them, the finished fries are spongy and dull. McDonald's fries may arrive at their restaurants frozen, but as even Eric Schlosser admits, in terms of flavorfulness, balance of sugar and starch content, and "mouth feel," the crisp fries that come out of McDonald's fryers are hard to beat.[45]

Reformers like Schlosser rightly praise In-N-Out for paying part-time workers a few dollars an hour above the minimum wage and providing full-timers with medical, dental, vision, and life insurance. In those regards, In-N-Out's management breaks ranks with the rest of the restaurant industry, whose lobbying arm, the National Restaurant Association, tirelessly opposes legislation to increase the minimum wage or push employers to provide decent benefits. Whether In-N-Out employees are happier or better treated than McDonald's workers overall, however, is an open question. So far as I have been able to determine, comparative studies of workers' satisfaction do not exist, and the only differences I have detected through casual observations at the branches near my neighborhood are the relative absence of adults and downtime at In-N-Out. At the McDonald's locations, perhaps half of the employees are post-adolescents and the pace is quick but not frenetic; by contrast, nearly everyone at these In-N-Outs appears to be in his or her mid- to late teens and working at fever pitch.

Both companies insist that their restaurants are wonderful places to work. In its annual reports to shareholders, McDonald's brags about its inclusion on independent rankings such as "Best Employers for Working Mothers in the U.S." and "America's Best Companies for Minorities." And even as In-N-Out describes it work environment as "fast paced and fun," the first item in McDonald's vision statement has the company calling on itself to "be the best employer for our people in every market around the world."

The reality does not match the hype, of course. Sociologist Jennifer Talwar's study shows that McDonald's and Burger King, another fast-food giant, have a long way to go before either becomes a model employer. Working at those chains' restaurants and interviewing fellow employees, Talwar learned firsthand how exploitative their practices can be. She describes how, to minimize labor costs during slow periods, managers sent workers home or on extended breaks, or canceled their shifts entirely, without compensation. To avoid paying unemployment insurance for employees they sought to fire, managers drove them to quit by drastically cutting their hours or scheduling them to work the least desirable shifts.[46]

Nor do experience and hard work consistently pay off. Fast-food companies do a great job of publicizing rags-to-riches stories about people like Phil Hagans, an African American from one of Houston's poorest neighborhoods whom the *Wall Street Journal* profiled as the burger flipper who then owned four McDonald's and a Mercedes. On its Web site, Burger King tells a similar story about Carlos Motes, a Cuban immigrant who rose through the ranks at Burger King from the broiler to restaurant manager to regional director. Far more common in McDonald's and Burger King, as Talwar shows, are employees with years of experience who remain at or near poverty wages and in fear of being laid off. The typical crew member's likelihood of becoming a rich franchisee or company executive is not much higher than the chance of winning a jackpot in Atlantic City.[47]

When Ray Kroc wrote his autobiography three decades ago, he could pass along without embarrassment the contention of a McDonald's franchisee that he "has made millionaires of more men than any other person in history." A substantial number of franchisees—including the McDonald's owner I interviewed—did indeed come from modest backgrounds. The first franchises Kroc sold in the mid-1950s went for $950. By the late 1960s, when this owner got into the business, McDonald's was a publicly traded company with more than a thousand locations, but a franchise still cost under $100,000.[48]

Today you need to have several hundred thousand dollars in cash and ready access to much more just to *apply* for a franchise at McDonald's or Burger King. But at least McDonald's and Burger King continue to offer people who can pull together the money a chance to own their own businesses. Practically no other major multinational corporation in any industry can say that, and neither can family-owned restaurant chains like In-N-Out.

The big fast-food companies afforded clear advantages as well for some of the low-income immigrant workers Talwar got to know, who used jobs in those places to learn English or to move beyond the limited work opportunities in the ethnic enclaves where they lived. Real advancement may be scarce at fast-food restaurants, but as Talwar came to appreciate, it is scarcer still in other lines of employment open to her co-workers, such as driving taxis or laboring in the underground economy.[49]

She described the working conditions of a Taiwanese immigrant employed as a crew member at a McDonald's in Chinatown. The woman's husband, a cook at a Chinese restaurant in Chinatown owned and staffed by Chinese immigrants, has "a brutal schedule of twelve to sixteen hours a day, six days a week, and compensation at far below the minimum wage, has no time, energy, or resources to pursue another job, or even to take English-language classes."[50]

No one ought to romanticize the wife's job at McDonald's, but I recoil when I hear self-described progressives condemn fast-food chains for their labor practices even as they frequent "authentic" ethnic restaurants where the employees "are likely to work in slave-like conditions," as a *New York Daily News* story put it.[51]

Similar Work, Different Food

A remedy that Schlosser and other reformers suggest to alleviate the ill treatment of fast-food workers—unionization—is almost unthinkable at ethnic dives, or for that matter, almost anywhere else that serves what the culinarily correct regard as cool or safe. Either these places are too small for unions to be interested, or they posture themselves as so caring that their workers would never want or need to organize.[52]

Consider O'Naturals, a small chain with restaurants in New England, whose founder, Gary Hirshberg, told me his goal is to become the "natural and organic alternative to McDonald's," with outlets near every McDonald's and Wendy's in the country. O'Naturals pays somewhat better than those conventional chains, but for the extra bucks, employees are expected to do ideological as well as physical labor. The company's job application specifies that even kitchen assistants must "become passionate around the recycling program of O'Naturals."[53]

To hear Hirshberg and Mac McCabe, the CEO of O'Naturals, tell it, working at their restaurants is heavenly. "We live off the byword that great customer experience comes from great employee experience, and we put a lot of value on that and really embrace them. We offer them, if they're willing to stay with us, room to grow, including financially. And we make it a hip, happy, fun place to work," McCabe told me.

Listening to McCabe, I had a feeling of déjà vu, as if I had heard this claim before, and in pretty much the same language. Subsequently I discovered that in fact I *had*—in recruitment materials from McDonald's and in a comment by one of McDonald's more famous former employees. "We can only provide the best customer experience when we provide the best employee experience," McDonald's proclaims. And Laurie Anderson, the performance artist, after she worked for a couple of weeks behind the counter at a McDonald's in Manhattan in preparation for a piece she was creating in 2002, told a reporter, "It was absolutely the opposite of what I expected. People who worked there were—and this is no joke—genuinely happy. Not to say anything about McDonald's in general, but this one has that feeling of camaraderie and fun. We were proud of what we were doing. We were fast. We joked around all the time."[54]

In arguing for their superiority, restaurants like O'Naturals draw sharp contrasts between their companies and their competitors, when really, the picture is considerably grayer. The way the O'Naturals folks tell it, McDonald's employees are miserable drones with nothing to be proud about. "When you really look at what the expectation is of the employee at a mainstream fast-food place," McCabe maintains, "it is that they're not capable of doing anything. It's basically reheating this frozen food or pushing buttons at a cash register. It's a dumbing down."

With such demeaning stereotypes, the prospects of fast-food workers' gaining respect in the larger society—or a living wage—seem slim. In actuality, jobs at the big fast-food chains tend to be appreciably more demanding and less robotic than outsiders suppose. "The universal aim of McDonald's and Burger King," Talwar notes in her study, "is for each employee to become proficient in every station and auxiliary task so that workers are interchangeable." These chains strive to create employees like the man who trained Talwar at a Burger King where she worked. "He was an impressive sight when he kept an eye on three or four monitors at once and darted from one station to another, his hands moving like a magician's."[55]

To my eye, the main points of contrast between O'Naturals and McDonald's or Burger King lie less in the types of work the employees do than in the kinds of food they sell and who buys it. Gary Hirshberg tells reporters he wants "to emulate the fast food places in every respect," but in point of fact, O'Naturals doesn't sell burgers and fries. It sells "herb roasted heirloom potatoes" and flatbread sandwiches with a choice of salmon, chicken breast, prairie-raised buffalo meat, slow-roasted beef, or a vegetable patty. Some of the most popular items on the menu are stir-fry noodle dishes.[56]

Hamburgers *were* on the menu when O'Naturals first opened in 2000, but as McCabe, a Harvard MBA, explained, they didn't sell. "It came down to a business decision. If there isn't much demand for burgers, then why are we keeping our grill going eleven hours a day burning propane? Why do we have a second prep chef during the peaks just to deal with hamburgers?"

Far from being typical McDonald's patrons, O'Naturals' target market is, in McCabe's words, "the natural-foods-supermarket family," people who feel reasonably good about what they prepare at home, but while chauffeuring their offspring "are always making this decision they despise. Between violin and soccer practice, they are taking themselves and their kids to places they hate," says McCabe, who told me he never eats at mainstream fast-food places. ("I hate the food. I mean, I *really* hate all that stuff coming out of the Frialator. It doesn't do a thing for me.")

McCabe described O'Naturals' core customers as "cultural creatives," a term marketers use for consumers who distrust large corporations and are willing to pay more for brands they consider honest, authentic, wholesome, and environmentally responsible. (A flatbread sandwich at O'Naturals costs about twice as much as a Big Mac.) The prototypical cultural creative is an upper-middle-class woman who listens to National Public Radio and believes in holistic medicine.[57]

If he is serious about competing with the conventional burger chains, it's hard to imagine how Hirshberg, O'Naturals' founder, whose other business is the yogurt company Stonyfield Farms, can succeed with a strategy of catering to a clientele so unlike the average fast-food purchaser. As of this writing, all of O'Naturals' locations are in tony towns and suburbs. Stonyfield Farms, by contrast, racks up annual sales of $150 million selling four hundred thousand cases of yogurt a week nationwide, most of them in dairy sections of conventional grocery stores.

Unlike the menu items at O'Naturals, the products from Stonyfield Farms look, taste, and are priced much like their conventional counterparts. Were Hirshberg to run the yogurt company on the O'Naturals model, the products would be made with soy instead of milk, and they would be sold at a premium solely at select Whole Foods locations.

Can a Burger Be Pure?

A profitable, safe-treyf alternative to the traditional fast-food burger has yet to be invented. The big guys have had no greater success than the upstarts. Over the past couple of decades, McDonald's and Burger King have invested multimillions in research, development, and marketing, with little to no success. The McLean Deluxe, introduced with much fanfare in 1990 and made of beef mixed with seaweed, had two-thirds less fat than the company's regular burgers, and no discernible following. McDonald's quietly dropped it from the menu in 1996.

More recently, rather than water down the beef, the top chains have tried eliminating it altogether, with no greater success. McDonald's developed a veggie burger and test-marketed it in some New York and California locations, but franchisees nationwide did not pick it up. And Burger King's BK Veggie Burger, on the menu nationally since 2002, has sold slowly and in unexpected ways. The product-development team had envisioned its audience as people seeking healthier fare, who would make a meal of the veggie burger plus a green salad and bottled water. "But when we put that in front of consumers," recalls Peter Gibbons, the team's leader, "they were like, 'Hell no! I'm not doing that! If I eat a veggie burger, that means I can come back one time more often or I can upsize the fries because I've made a concession with the veggie burger.' What we found was that people didn't want to avoid eating beef, they didn't want to do anything of the sort, they just considered this to be permission food."

Counterfeit burgers satisfy the bloodlust of neither side in the burger wars. To hamburger fans, anything other than the real deal (a beef patty with cheese on a fully dressed bun) is at best an appetizer. And to the anti–fast food crowd, fake meat is a bogus solution. "The mere absence of meat and cheese from the BK Veggie says nothing about its nutritional value. Froot Loops, Pepsi, and Burger King's own French fries, for that matter, are also free of animal products, but few health advocates would seriously recommend consuming these foods as part of a well-balanced meal plan," wrote Rich Ganis, a director of the Center for Informed Food Choices, in an op-ed in the *Los Angeles Times* soon after the Burger King product came out. "Promoters of the BK Veggie are doing the public a serious disservice by suggesting that it is anything other than a highly processed, nutritionally deficient junk food that just happens to be meatless."[58]

One might naively have expected that a group like Ganis's, which advocates vegan diets, would celebrate a low-fat, lower-calorie burger that, when ordered without mayonnaise, is devoid of animal products. The ingredients that account for 99 percent of the patty read like the contents of a vegan's pantry:

mushrooms, water chestnuts, brown rice, rolled oats, onions, corn oil, carrots, green peppers, red peppers, black olives, salt, pepper, and basil. But for someone of Ganis's sensibilities, the presence of phrases like "grill flavor" and "hydrolyzed corn gluten" on the ingredient list for the remaining 1 percent eclipses all of that.

Echoing a sentiment voiced frequently by vilifiers of fast food, Ganis concluded his op-ed with a rejection of what he called "the industrial food economy." Reformers should accept nothing less than "a food system that provides people with produce in its whole, unadulterated form, as nature meant for it to be eaten," he wrote.

Far be it from me to question those who know what nature intends. Sometimes I have trouble making out the intentions of my fellow humans, never mind the cosmos. Still, I can't help but wonder how hundreds of millions of people have enjoyed fast-food burgers and lived to tell the tale if Mother Nature had entirely different plans. Until the naysayers come up with an equally tempting and affordable alternative, maybe they should hold off pontificating about what may well be the most widely consumed entrée in history.

7

What Made America Fat?

Hint: It's Not Just the Food

But how about obesity? Whatever else can be said about McDonald's and its competitors, can anyone doubt they are largely responsible for Americans' getting larger over the past quarter century—a lot larger? Throughout the 1960s and 1970s, about 13 to 15 percent of Americans were obese, a number that more than doubled between 1980 and 2000.* The rate of increase for those who became merely overweight was much less, but add the numbers together and you get the shocking statistic that nearly two-thirds of Americans weigh more than health officials recommend.[1]

Innumerable observers have echoed Eric Schlosser's observa-

---

*"Obesity" is defined as a body mass index (BMI) greater than 30, where BMI is computed as weight in kilograms divided by the square of height in meters. "Overweight" is defined as a BMI greater than 25. A five-foot eight-inch person is officially overweight at 165 pounds and obese at 197 pounds.

tion in *Fast Food Nation* that the proportion of fat Americans soared during the same period as the rise of the fast-food industry. Like a hundred other taken-for-granted truths about Americans' newfound girth, however, this seemingly obvious connection falls apart on closer scrutiny.

In fact, the explosion of the fast-food industry predated the upsurge in obesity. It was in 1966 that signs outside McDonald's restaurants boasted "over two billion sold." By the mid-1970s, McDonald's had became nearly a billion-dollar-a-year business with more than ten thousand locations nationwide. Ray Kroc's main competitors had as many more. Yet obesity rates barely budged during the 1960s and 1970s.[2]

The list of explanations for what got us fat over the past quarter century, each with its own ring of truth and band of devoted scientists, activists, and dieters, is longer than a well-stocked smorgasbord. Journalists and government officials typically favor what I call the "fiscal model," which holds that "energy is deposited by eating food, that exercise and metabolism withdraw it, and that body fat is a sort of corporeal balance sheet," as S. Bryn Austin, an instructor at the Harvard School of Medicine and critic of the model, summarized. A version of the gospel of naught, the fiscal model blames the obesity epidemic on overeating and inactivity. As a writer for *U.S. News & World Report* put it, "Overweight results from one thing: eating more food than one burns in physical activity."[3]

Believers in the fiscal model contend that in the absence of additional exercise, it took no more than a few extra bites or slurps a day by most Americans to produce the obesity epidemic. "To gain 15 pounds in a year, you only have to have an imbalance of 150 calories a day, which is one soft drink," Dr. Thomas Robinson, an obesity researcher at Stanford, told a *New*

York Times reporter. "Even a Life Saver is 10 calories. An extra Life Saver a day is a pound a year."[4]

In this view, calories are like germs. Proponents of the fiscal model speak of calories lurking in unexpected places and finding their way into our bodies when we're scarcely aware. The notion dates back to 1918, when Lulu Hunt Peters published *Diet and Health with Key to the Calories*, America's first diet bestseller. The book brought the concept of the calorie to the general public. "You should know and also use the word calorie as frequently, or more frequently, than you use the words foot, yard, quart, gallon and so forth," Peters instructed. "Hereafter you are going to eat calories of food. Instead of saying one slice of bread, or a piece of pie, you will say 100 calories of bread, 350 calories of pie."[5]

Generations of Americans have followed her command, aided, during much of the period when our collective weight shot up, by federally mandated labeling of the calorie content of every packaged food product.

A Red Herring

But could it be, as some scientists and diet gurus maintain, that the fiscal model radically oversimplifies the process of weight gain and loss? Might calories not be the true culprits after all?

Some argue it is the types of foods Americans eat that have made us fat. "Fat makes you fat," the diet guru of the 1990s, Susan Powter, famously proclaimed, and Dr. Dean Ornish, of *Eat More, Weigh Less* fame, continues to preach that gospel. Other diet docs, carrying forward the teachings of the late Robert Atkins, insist it's the carbohydrates. And some, like the South Beach Diet mogul Arthur Agatston, split the difference.

Rather than excommunicate either food group entirely, they banish only those they deem "bad fats" and "bad carbs."

Or maybe the emphasis on food is itself misguided. "There is no evidence that fat people consistently eat more than the lean," William Bennett, a Harvard Medical School physician and longtime editor of the *Harvard Medical School Health Letter,* reported in 1982. In a book he cowrote that year with Joel Gurin, editor of *American Health* magazine, and in articles in medical journals over the next dozen years, Bennett showed that you cannot predict people's weight gain by how much they eat.[6]

"Food is a red herring," he wrote. "It is perfectly possible for some people to eat a lot and gain very little, whereas others gain weight while eating abstemiously." Armed with experiments showing that fat people consume no more calories than thin people, Bennett described the fiscal model as fatally flawed—a conclusion supported by later studies that compared the diets of men and women across a wide weight range, and by studies of twins. In these latter experiments, scientists fed pairs of identical twins many more calories than they customarily ate or, conversely, put them on an exercise regimen to "burn off" calories. After a few weeks, there was great variation *between* the pairs of twins but hardly any *within* each pair. Unrelated individuals gained or lost widely different amounts, while differences between twin siblings were minimal. This implies that people's weight is governed more by their genes than by how many calories they eat or deplete.[7]

Those who prefer to blame the obesity epidemic on food and sloth consider it absurd to propose that genes may be an important culprit. The obesity epidemic materialized over a couple of decades, whereas genes take at least a couple of *generations* to change, they correctly note, and from those facts, they wrongly infer that the epidemic must have resulted from America's "food-rich, activity-poor environment" and "a certain sin known as gluttony, which has somehow gotten a good name," as author Greg Critser says in his book, *Fat Land: How Americans Became the Fattest People in the World.*[8]

To the extent that devotees of the fiscal model grant any role to inheritance, they favor the so-called thrifty gene hypothesis. Because our ancestors frequently faced food shortages and famine, that story goes, we evolved to eat and store energy. In an environment of easy access to cheap and appetizing calories, we're programmed to gobble up and retain more than we need. The real surprise is that anyone stays thin in such an environment.

Twin studies and actual patterns of obesity in the U.S. tell a different tale. There seems to be no species-wide tendency; rather, only a relatively small minority of people appear to be disposed to obesity. Jeffrey Friedman, a prominent obesity researcher at Rockefeller University, has shown that the obesity rate shot up not as a result of big increases in weight throughout the population, but rather because of a threshold effect. A sufficient number of Americans were just below the cutoff for what officially qualifies as obese. By gaining a modest amount of weight, they crossed the threshold and got reclassified from "overweight" to "obese." Although the obesity rate increased by a whopping 30 percent between 1991 and 2001, for example, the typical American gained less than a pound a year. But that fairly modest weight gain was enough to push substantial numbers over the threshold from "overweight" to "obese."

Friedman notes that many Americans added little or no weight during the obesity epidemic. Only the very obese added twenty-five pounds or more, and different ethnic groups gained different amounts of weight. These facts strongly suggest, Fried-

man argues, that a subgroup within the population is genetically predisposed to obesity and another subgroup is not. The
two subgroups may have different genetic lineages, he contends.
The portion of the U.S. population whose ancestors resided in
the Fertile Crescent and parts of Europe where a favorable cli-
mate or domestication of plants and animals made food shortages less of a problem may actually have inherited a *resistance* to
obesity. "Might it be," Friedman asks, "that it is the obese who
carry the 'hunter-gatherer' genes and the lean that carry the
'Fertile Crescent' or 'Western' genes?"

The principal point of natural selection is to ensure repro-
duction, after all, and obesity increases the likelihood of mis-
carriage. So as Friedman suggests, "where the risk of starvation
is reduced, one might expect genes that resist obesity and its
complications to have a selective advantage."[9]

The Law of Unintended Consequences

Or maybe neither bad genes nor Big Macs are the right place to
look for the causes of the obesity epidemic. Another large body
of evidence points in a different direction, to changes in the
American economy. During the decades when Americans'
weight shot up, so did levels of economic hardship and insecurity. In the 1980s and 1990s, more Americans lost their jobs
than at any time since the Great Depression, and those who did
have jobs worked longer hours. About a third of the population
became poorer during this period, and millions more had difficulties maintaining their lifestyles because the raises they received did not keep up with inflation.[10]

Who suffered the most from these misfortunes? The same
sectors of the population who gained the most weight: low-
income Americans and ethnic minorities. The wealthiest 20

percent of Americans—who now control about 80 percent of the nation's total wealth—have relatively low rates of obesity. So do those whose socioeconomic status has improved.[11]

A key link between the obesity epidemic and economic hardship is chronic stress. Stress provokes the body to produce less growth hormone, a substance that reduces fat deposits and speeds up metabolism, and *more* of what are called stress hormones, which provoke cravings for soothing substances like glazed doughnuts and chocolate fudge ice cream.[12]

People don't invariably respond to stress by gobbling comfort foods, however. Many opt instead for cigarettes, and therein lies a luscious little irony. The obesity epidemic that government agencies and advocacy groups are battling to reverse resulted in part from the success of antismoking campaigns by these same organizations in the recent past. The number of smokers declined by about a third during the 1980s and 1990s, and when people give up smoking, they tend to gain weight.[13]

We social scientists call this the law of unintended consequences. Roughly the sociological equivalent of Newton's third law, it holds that any social intervention that produces beneficial outcomes will be likely to give rise to unintended negative effects as well. The obesity epidemic cannot be explained entirely, though, by way of the law of unintended consequences. Even a valid application of the law, such as the connection between antismoking campaigns and obesity, accounts for only a fraction of the nation's added tonnage. (The ranks of the obese include people who never smoked, and some people give up smoking without getting fat.)[14]

Some attempts to apply the law of unintended consequences to the obesity epidemic amount to little more than camouflaged slander. Consider this explanation for childhood obesity from the head of a health advocacy and research organization, quoted in the *New York Times*: "In many households today, both parents work, so kids return to an empty house and settle in front of the television." As Natalie Boero, a doctoral student at Berkeley who studies obesity, notes, such observations implicitly blame mothers, "whose paid work is often seen as secondary or unnecessary."[15]

In this view, many younger victims of the obesity epidemic have, if not their own working mothers to blame, the social movements of the 1960s and 1970s that encouraged women to have jobs and careers outside the home. A variant of the fiscal model, this theory paints a picture of neglected kids downing Doritos in front of idiot boxes when they ought to be out playing sports, cheered on by their moms. But research on families in which both parents work refutes such shopworn stereotypes of so-called latchkey kids and their negligent parents. Studies document that children get at least as much time and care from their working moms today as earlier generations got from stay-at-home moms.[16]

Women continue to be blamed by obesity theorists. Even as present-day moms are denounced for having jobs, an earlier generation of women gets impugned for having participated in a popular trend of the 1950s and 1960s: bottle-feeding. Noting that millions of today's obese adults were babies during the period when breast-feeding rates took a dive, some argue that "obesity is a result of inadequate breast feeding."

Undurti Das, a widely published medical researcher, made that claim in the research journal *Nutrition*, and he offered a scientific explanation for how a diet of infant formula might set people up for obesity later in life. The physiological link, he proposed, is an ingredient in breast milk that's absent from infant formula: long-chain polyunsaturated fatty acids, or LCP-UFAs. Without enough LCPUFAs, insulin receptors in the brain malfunction, provoking hunger and the storage of calories as fat, and over time, Das hypothesized, obesity.

It's an explanation that jibes with a body of scientific evidence on the biological pathways to obesity, but whether the connection Das set out to explain is real is another matter entirely. Fewer mothers of baby boomers may have breast-fed, but the pertinent question is, Do bottle-fed babies more often become obese adults than their breast-fed peers? Apparently, they do not. Studies find roughly equal rates of obesity among groups of adults who had been breast-fed and those who had been bottle-fed. And Americans now in their teens and twenties were weaned at a time when rates of breast-feeding rose significantly, yet they have high rates of obesity.[17]

Why Fast Food Takes the Fall

With all the social changes that have taken place during the lifetimes of present-day Americans, and scores of conflicting findings in the scientific literature, there are scapegoats for the obesity epidemic to fit every personal and political predilection. Members of the Traditional Values Coalition can accuse working moms. Members of La Leche League can point to makers of infant formula.

And anyone who dislikes fast food can go after that industry. According to a standard explanation for both the obesity epidemic and its concentration among the lower classes, "America's least well-off are so surrounded by double cheeseburgers, chicken buckets, extra-large pizzas and supersized fries that they are more likely to be overweight than the population as a whole" (Gregg Easterbrook in the *New York Times*).[18]

Even putting aside the broader failings of the fiscal model from which it is derived, and the problems of chronology I mentioned at the outset, the fast-food theory has little to commend it. As a prominent obesity researcher at the U.S. Department of Health and Human Services who asked to remain anonymous told me, "There's a lot of subtle and not so subtle bias. From going to all these talks about the obesity epidemic, you would think that McDonald's and other places where the 'wrong' sort of lower-class people eat are calorie-dripping hellholes, and expensive classy restaurants serve only fat-free vegetables and no desserts.

"No one ever uses Starbucks as an example, but a Frappuccino is as oversized and calorie-laden as anything McDonald's can dream up. But the person giving the talk probably goes to Starbucks him- or herself and wouldn't be caught dead at McDonald's."

Only a small number of studies have attempted to test the fast-food hypothesis directly, and they have come up with mixed results. Contrary to the impression given by some journalists and activists that dining in a fast-food restaurant "is like sitting in a room set up by aliens from another planet to fatten us up before they eat us" (Gersh Kuntzman in *Newsweek*), some studies find no association between people's body weight and whether they eat in fast-food restaurants.[19]

What's more, some of the studies cited by advocates of the fast-food theory do not actually support it. Take this assertion from a paper in the *Journal of the American Medical Association* in 2004: "The increase in fast food consumption parallels the escalating obesity epidemic, raising the possibility that these two trends are causally related." When I dug up the only evidence the authors cite in support of that dubious claim—a paper published two years earlier—I discovered that it said no such thing. (Concerned that I had somehow missed something,

I e-mailed one of the authors of the cited paper. "You're right about our article. We don't say anything about fast-food consumption," she replied.)[20]

Or take the study of nearly five thousand adolescents that Greg Critser relies upon in *Fat Land* to support his condemnation of fast food as a perpetrator of obesity. Critser uses words like "striking" and "amazing" to describe what he saw as the study's findings, but when I read the study itself, I found something more striking than the finding Critser cites. The researchers did report, as Critser highlights, that adolescent boys who eat fast food consume more calories than boys who never visit fast-food places. And they speculated that adolescents who develop a fondness for fast food may be at greater risk of obesity later in life. But the findings of their study do not show that fast food causes obesity. Quite the opposite. "In the present study, no association was observed between frequent fast food restaurant use and obesity, even though fast food restaurant use was significantly positively associated with energy and fat intake," write the University of Minnesota epidemiologists in the *International Journal of Obesity*.[21]

Far from finding that teens who eat fast food are fatter, they determined that boys who dine on fast food three or more times a week weigh significantly *less* than those who eat there less frequently. And teens from families that eat together consume fast food about as often as those from households that seldom have meals together, the Minnesota researchers found, contrary to popular belief.

The Nostalgia Trap

The puzzle of what caused and what continues to cause Americans to put on pounds is not likely to be solved by pining for a time when families ate together. A typical dinner menu in a typical American home during the decades prior to the obesity epidemic bears no resemblance to any of the versions of the gospel of naught presently advanced as antidotes for overweight. Loaded with calories from saturated fat, family meals of the 1950s through 1970s also didn't scrimp on carbs. As economist Todd Buchholz recalls, "Meat loaf, fried chicken, butter-whipped potatoes, and a tall glass of whole milk may have kept us warm on a cold winter evening, but such a diet would surely fail a modern test for healthy living. And let's not even discuss a crusty apple pie or bread pudding for dessert."[22]

How conveniently we forget that this was an era when parents and grandparents, who had survived the Great Depression and a world war or two, encouraged their kids to finish second and third helpings because, as some of us were repeatedly reminded, "children are starving in India." Parents of yesteryear also liked to haul their brood to all-you-can-eat buffet restaurants and church socials where the assembled multitudes downed quantities of food that rivaled Morgan Spurlock's grossest pig-outs in *Super Size Me*.

Speaking at an "obesity summit" presented by *Time* magazine and ABC News, where he was introduced as a hero who lost 105 pounds, Mike Huckabee, the governor of Arkansas, attributed his obesity as an adult to having "grown up eating all the wrong things." "If you really, really want to get against a southerner, try to take his guns or his gravy away, and God help you," Huckabee joshed, and told a joke about his grade school teacher asking the kids to bring a symbol of their religious faith for show-and-tell. "A Jewish boy brought a menorah, a Catholic girl brought a rosary. I brought a covered dish."[23]

The line got a knowing laugh from the audience, but Huckabee neglected to explain why the obesity rate was so much lower back when he and his neighbors were downing stacks of buckwheat pancakes in syrup and sides of grits; chicken-fried steaks with fried okra; and the pecan, coconut cream, chocolate meringue, and sweet potato pies that earned Arkansans a reputation as a pie-mad people.[24]

Go on a Diet, Gain Weight

Here's a paradoxical possibility. Maybe the nation's obesity statistics were swelled not mostly by rich eats, but by *avoidance* of such foods. A great deal of evidence points in that direction, not the least of which are studies that find a strong correlation between dieting and being obese. Not merely a reflection of the fact that lots of overweight people are on diets, the correlation signals that, as a group of Harvard and Stanford researchers observed, for substantial numbers of people, "dieting to control weight is not only ineffective, it may actually promote weight gain."[25]

The researchers made that assertion in a paper in the journal *Pediatrics*, where they reported the results of a study of almost fifteen thousand boys and girls between the ages of nine and seventeen. Over the three years the researchers followed the kids, the dieters in the sample gained more weight than the nondieters and were more likely to engage in binge eating.

Other research also suggests that dieters and those who encourage dieting bear more than a little responsibility for the obesity epidemic. For example, the University of Minnesota team who conducted the fast-food study found in a separate survey that adolescent boys whose mothers encouraged them to diet were significantly heavier than their peers whose mothers left them alone about their weight, and seven times more likely to engage in binge eating.[26]

Another set of studies finds young girls are more likely to become overweight or obese later in life if their parents put them on diets by restricting them from eating "junk" foods.[27]

In surveys of obese adults, social scientists at the University of California came upon further evidence of the fattening effect of dieting. Instead of looking at young dieters to see if they were more likely to get fat, these researchers asked obese adult women when they had started dieting. Nearly two-thirds had gone on their first diet before age fourteen, and the heaviest women in the study had dieted earlier and more often than the rest.[28]

"For this group of women, dieting has actually promoted their obesity," Joanne Ikeda, the lead researcher on that study and former director of the Center for Weight and Health at the University of California, Berkeley, reported. Noting that the number of obese and overweight Americans increased at the same time that dieting became commonplace, Ikeda made the obvious point: "One has to wonder, is there not a link between the two?"[29]

In *Losing It*, her exposé of the $50 billion diet industry, author Laura Fraser identified the likely link. Bingeing. "When we starve ourselves, our bodies call out for help with hunger pangs and cravings, and our minds plot a rebellion," she wrote. "Diet foods, in particular, make us want to overeat. When we eat diet foods, they're usually a cheap substitute for 'bad' foods, and we aren't really fooled. We end up overcompensating for our desires, eating more of the diet food than we should, looking for satisfaction."[30]

Lose Weight, Become a Hermit

By denying ourselves high-fat foods, high-carb foods, fast foods—or whatever the prevailing diet orthodoxy prohibits—

we may be doing little to bring down the country's gross tonnage. Rather than give up pleasurable foods, maybe we ought to forsake other things that studies have shown to be associated with weight gain: going to church, for example, or dining in groups.

Researchers at Brown, Purdue, and Cornell universities discovered that church members are more likely to be overweight than the rest of us, and the most religious Americans have especially high rates of obesity. Although social scientists who study the matter have not nailed down the link between religiosity and fat, they have some likely candidates. They note, for instance, that people who are more religious have lower rates of smoking.[31]

Active churchgoers have also been known to frequent those church socials I mentioned before, where in addition to lots of food, there are lots of eaters, a known risk factor for obesity. The more people present at a meal, the more they tend to eat, studies find.[32]

There may be another reason to stay away from crowds as well. Research by a group of scientists at Wayne State University suggests obesity may be caused by a coldlike virus called adenovirus-36.

These biomedical researchers are serious. The idea of "catching obesity" may sound like the premise for a *Saturday Night Live* skit, but the Wayne State scientists have found that overweight people are four to six times more likely to have the adenovirus than leaner folks. What's more, when they inoculated chickens, monkeys, and mice with an adenovirus, the animals gained weight and body fat without eating more.

The researchers contend that the spread of this virus or its cousins may explain why rates of obesity have been rising throughout the world in recent years. They point out that obesity would not be the first condition long thought to result from a bad diet and lifestyle that turns out to be caused by microbes. Ulcers, now understood to be caused by a bacterium, are an example, and pathogens have been implicated in heart disease as well.[33]

America's Number Two Killer?

The deeper I burrowed through the theories of obesity, the more of them I found.

I also found myself entertaining an irreverent thought. How much does it really matter that Americans are getting fatter? To the diet industry, it matters a great deal, but the profit motive aside, does plumpness really deserve all the attention and resources we devote to it?

Many of the theories of obesity are fascinating, to be sure, but with nearly two-thirds of Americans overweight or obese, their body types are now the norm, and medical science does not usually concern itself with trying to understand and prevent what has become the norm. Americans have also grown taller in the recent past (about four inches on average since the late 1800s), but little attention is devoted to that change, which conventional wisdom also attributes to changes in diet. Corporations and government agencies do not push pills, programs, or special menus for height control, even though tall people are more prone to an array of ailments, from orthopedic maladies to several types of cancer.[34]

Government officials and journalists justify their fixation on weight with debatable claims about the lethality of fatness. "Obesity on Track as Number One Killer," read the front-page headline on a *USA Today* story in 2004. Reporting on a study in the *Journal of the American Medical Association* (*JAMA*) that attributed 385,000 deaths a year to overweight and obesity, the story concluded with a quote from Julie Gerberding, one of the study's authors and director of the Centers for Disease Control. Americans need to learn to eat "healthy foods in healthy portion sizes and find ways to incorporate exercise into their everyday lives," she admonished.[35]

That Americans had been getting *more* exercise and eating more fruits and vegetables and less fat during the period when obesity rates shot up apparently didn't faze Gerberding. But some of the top obesity researchers within her own agency fumed when they saw the *JAMA* article blaming obesity and hundreds of thousands of resultant deaths on "poor diet and physical inactivity."[36]

"Everyone is really angry right now; this is a big scandal," a prominent researcher at the CDC told me soon after the article came out. She and some of her colleagues don't buy the claim that obesity and overweight are top killers. "A lot of the research centers at CDC wouldn't approve the article," she said. "There were protests about the methodology and the data, but the authors ignored all input from everyone else at CDC and went ahead with this stuff. Their argument doesn't even make sense, because you can be fat and have a good diet and be physically active, and you can be thin and have a poor diet and be sedentary."[37]

I'll refer to this scientist as Dr. Diver. An eminent obesity researcher whose own papers have appeared in top medical journals and are cited hundreds of times in the obesity literature, she spoke with me on the condition that I not use her name. (Government researchers are prohibited from making public statements that contradict their agency's official position.)

Diver called to my attention a couple of basic facts that undermine the idea that obesity is a major killer. Life expectancy has increased during the obesity epidemic. And most people die old. Three-quarters of people who die are over sixty-five, and 40 percent of deaths are people over eighty, Diver pointed out. "So the important question about the effect of obesity on death rates," she said, "is in old people, and the evidence suggests obesity is almost irrelevant there."

Highly critical of studies purporting to show that young people sacrifice years of their lives by being fat, Diver directed me to another study published in *JAMA*: "Years of Life Lost Due to Obesity." This study declared that highly obese white men in their twenties will lose thirteen years of life because of their weight—an eye-catching claim that was picked up by the news media and has been cited in numerous other journal articles. But when Diver examined the survey on which it was based, she came upon something curious. "I looked at that paper and said to myself, 'How many guys like this could there be in that data set?' I know those data sets well, and the answer is, there aren't any. There are no white males between twenty and twenty-nine with BMI's that high in those surveys. The finding is bogus."[38]

Lacking actual people in that age group and weight range, Diver explained, the researchers based their conclusion on hypothetical models. When Diver herself looked at the same data (the government's National Health and Nutrition Examination Surveys) she discovered that, in fact, a large proportion of healthy people are overweight.

Far from being the only authority who disputes the party line about the causes and consequences of weight gain, Diver is one of a dozen well-informed skeptics I came upon: physicians at major medical associations, research scientists with government agencies, and professors at leading universities. They all concur with an epidemiologist I interviewed from another division of the Department of Health and Human Services. "The evidence is just not there," he said, "to support the claim that if a healthy person maintains his weight at a so-called normal level all his life he will add years to his life span."[39]

Indeed, in a paper published in *JAMA* in 2005, four preeminent statisticians from the CDC and the National Cancer Institute showed that people officially classified as overweight actually have *lower* death rates than "normal" and "underweight" people. Using nationally representative data collected from 1971 through 2002, they established that while the truly obese do have higher death rates, the number of deaths attributable to obesity is about 112,000, less than one-third the number the CDC chief had been publicizing.[40]

The Weight-Centered View of the Universe

Nor, say the skeptics, do thin folks protect themselves from many of the ailments that government officials and health columnists have attributed to corpulence. No one denies that weight is a factor in type 2 diabetes or that extremely obese people suffer serious health problems as a result of their weight. But the critics offer good reasons to question whether mildly overweight people lower their odds of heart disease or cancer by dieting.[41]

"We talk about all these so-called obesity-related diseases, but that's not really their major cause," Dr. Diver told me. "Some are age-related diseases, and for some, such as heart disease, obesity is just one risk factor among many risk factors and not necessarily the most important nor the cause. I call this the weight-centered view of the universe: weight causes all the problems. If you actually look at the data, you find that everybody who is overweight and has hypertension, for example, would probably still have it regardless of their weight, because it's an age-related condition."

When critics dare to raise those sorts of matters publicly, orthodox obesity researchers treat them dismissively. In 2004, after Paul Campos, a law professor at the University of Colorado, published his copiously documented book, *The Obesity Myth*, advocates of the reigning dogma responded like politicians. They ignored the challenger except when journalists compelled them to comment.[42]

It was in response to a request from *USA Today* that Walter Willett, the Harvard professor of nutrition, commented on *The Obesity Myth*, and his remarks could hardly have been more derisive. Rather than address Campos's arguments and evidence, Willett wrote him off as "one lawyer with no experience and no medical training."[43]

In response to a question from the reporter about whether being overweight leads to serious illness, Willett repeated the claim that "there's a strong relationship between extra body weight and heart disease," though, as Campos and other skeptics emphasize in their writings, the number of deaths due to heart disease plunged rather than rose during the obesity epidemic. The skeptics cite as well autopsy and angiography studies that directly examine people's hearts and report no association between body weight and heart disease.[44]

If Willett's side has convincing evidence to refute its critics on these points, I couldn't find it. After the *USA Today* piece came out, I e-mailed Willett for studies showing that extra weight increases the risk of heart disease. He replied not with a bibliography but with a citation of one of a pair of papers he published in 1995 about weight and heart disease, and a couple of caveats. "This is just one study," he wrote, "and there is a vast literature on this. Campos either can't read or is purposely deceiving his audience."[45]

My high school debate coach taught me to ignore ad hominem attacks, which he said are a sure sign of a weak case. So I disregarded Willett's characterization of his opponent and went in search of the vast literature to which he referred. And sure enough, I had no problem locating papers that assert a correlation between increasing weight and heart disease. Many were at pains to emphasize, however, that "the increased cardiovascular risk associated with obesity is applicable in only a minority of the obese subjects," as a group of Swedish researchers noted in a paper in 2002 in the *International Journal of Obesity*. In their twenty-three-year study of more than twenty-two thousand men, the scientists found that even though, as a group, overweight and obese men had more coronary incidents than other men, fully 90 percent had none. Nearly all of the added risk of heart disease among the obese is due to factors other than the weight itself, the researchers found, such as high cholesterol or hypertension.[46]

As near as I can tell, not a single published study demonstrates that heart disease among the overweight and moderately obese results from their heft rather than from other factors that contribute to obesity and heart disease, such as smoking, poverty, stress, genetic predisposition, physical activity, depression, and quality of medical care.

The papers to which Willett sent me directly certainly do not settle the matter. Products of his Nurses Health Study, they take into account few of the many pertinent factors. With a sample of women all employed in the same profession, 95 percent of them white, Willett and his research team have no way to assess some of the alternative explanations. They do acknowledge the importance of smoking, but that leads them to exclude smokers from their analysis, which leaves them with a relatively small sample. Of the 115,195 women in the study, only 184 were non-smokers who died of cardiovascular disease during the sixteen years of study.

In a paper published in the *New England Journal of Medicine* in 1995, Willett and his colleagues reported a 60 percent increase in the death rate from cardiovascular disease for nurses who were moderately overweight (BMI of 27 to 29) compared to slender nurses (BMI under 22). That sounds like an alarming number. But how many actual people does it represent? Willett's team divided the 184 nurses into seven separate subgroups by weight. Presumably, even the largest of the subgroups had no more than two or three dozen women. With numbers that small, it would take only a few extra deaths of women in the overweight subgroup to produce the seemingly shocking statistic.[47]

I don't know the exact number of nurses involved, because Willett wouldn't provide me with data to compute it. "Unfortunately, many of the detailed numbers you request are not in the published papers; editors are notoriously reluctant to include all the details we would like to publish," he replied to an e-mail in which I asked where I could find some of the information missing from his papers. When I followed up with a request for someone on his staff to pass along the numbers, he wrote back, "I would like to help you, but we don't have someone to do that; we have asked the National Institutes of Health for such funds."

Money Triumphs over Fat

Willett was equally unhelpful in response to my request for studies that support another of his assertions in the *USA Today* debate, that "many people manage to reduce their weight by careful diet and regular activity."

Paul Campos and other skeptics contend that regardless of who is right about whether being overweight causes heart disease,

Willett's side does a disservice by advising that anyone can slim down by cutting back on disapproved foods and becoming more active. In reality, say the critics, only a small proportion of people succeed in taking off much weight and keeping it off.

My review of the research literature supports that conclusion. In an article in the *New England Journal of Medicine* in 2002, for example, physicians from obesity programs within the National Institutes of Health report that people who devotedly diet, exercise, and get counseling for four to six weeks can expect to lose 5 to 10 percent of their weight. But echoing the conclusions of studies I mentioned earlier, they add: "For the vast majority of persons, weight loss is followed by a slow, inexorable climb to the preintervention body weight—or even higher."[48]

When I e-mailed Willett for evidence supporting his more optimistic conclusion, he had little to offer. "There is abundant evidence that many people do control their weight; for example, the rate of obesity in groups with higher education is only half that of low-education groups. These are the folks I see running along the Charles every morning," he wrote, as if one can generalize from genteel joggers in Cambridge to how other Americans would look if they ate less and ran more.[49]

College graduates may be thinner for any of a number of reasons, including, as we've seen, their genes and their insulation from chronic stress. Or maybe the lower average BMIs of America's elite indicate only that "in America, money triumphs over the most resistant fat, which eventually succumbs to regimens that only the very rich, or the fanatical, can afford," as the Cornell University literary critic Richard Klein has argued. Wealthier Americans certainly have more time and money to go to extreme lengths in pursuit of thinness—a pursuit that is as much about preserving their social position as protecting their health. In the past, when food was scarce, a big belly was a mark of wealth, status, and attractiveness, as is a flat stomach now that most of the population can afford double cheeseburgers.[50]

In the eyes of many, slenderness is a sign of discipline and healthiness, but skeptics reasonably question whether that self-serving presumption ought to be imposed on the entire population by way of ultimatums to shed pounds. Healthy people who lose weight do not live longer than those who do not, studies find. A high but stable weight is safer than repeated fluctuations, and research suggests that yo-yo dieting damages the immune system and may increase the risk of heart disease and type 2 diabetes.[51]

Blinded by the Lithe

In 1998, the physician editors of the *New England Journal of Medicine* wrote an editorial about obesity. Their overall conclusions accord with my reading of more recent studies as well. "Given the enormous social pressure to lose weight, one might suppose there is clear and overwhelming evidence of the risks of obesity and the benefits of weight loss. Unfortunately, the data linking overweight and death, as well as the data showing the beneficial effects of weight loss, are limited, fragmentary, and often ambiguous," Jerome Kassirer and Marcia Angell wrote.[52]

Lamenting the fact that "many Americans are sacrificing their appreciation of one of the great pleasures of life—eating," Kassirer and Angell rejected the fiscal model of obesity as simplistic and endorsed in its place a "set point" theory of body weight. In this view, each of us is genetically programmed to be within a particular weight range, and our central nervous system defends that set point by altering our appetite or metabolism if we try to drop or add much weight. The set point can be overridden temporarily by zealous dieting and exercise, but as the editors note, "when these extreme measures are discontinued, body weight generally returns to its preexisting level."[53]

Noting that "in this age of political correctness, it seems that obese people can be criticized with impunity, because the critics are merely trying to help them," Kassirer and Angell called on doctors to help end discrimination against overweight people. Their plea is warranted not only on humanitarian grounds (studies document persistent discrimination by teachers, college admissions officers, employers, and health care providers), but for medical reasons as well. A body of research shows that discrimination itself has negative effects on people's health.[54]

Talk about a vicious cycle: the stress, social isolation, and lower education and income levels engendered by discrimination leave people more prone to weight gain, and thus to more discrimination.[55]

Focused on diet and exercise, many researchers, journalists, and policy makers fail to take sufficient notice of this cycle. Their myopia blinds them to another important contributor to obesity and disease among low-income Americans as well: food insecurity. According to Greg Critser, lower-income Americans are fatter because, unlike the rich, "the more insightful and longer-living of whom have understood that the price of abundance is restraint," poor folks overeat and underexercise. But in reality, it is people who periodically face the prospect of not having *enough* to eat who have the highest rates of obesity.[56]

As it happens, neither part of Critser's statement is accurate. Wealthy Americans are far from restrained in their consumption patterns—witness their SUVs and oversize homes. And large-scale studies show that those who suffer from food insecurity are more likely to be overweight than others of the same age, ethnicity, and income level. Exactly how food shortages lead to obesity is hotly debated, but the fiscal model can't account for the seeming paradox. In a study where researchers compared people with similar diets and levels of physical activity, they found that those experiencing food insecurity were more apt to be overweight.[57]

Rather than sloth and high-calorie foods, the primary culprit in obesity among the food-insecure may be one we met earlier: binge eating. The eating patterns of people who periodically run low on food resemble those of dieters, albeit for different reasons. When their food stamps run out, their kids' medical expenses use up the food budget, or the local food bank shuts down, they eat very little. Then, once food becomes available again, they binge. Over time, their bodies adapt to the alternations by converting more of what they eat to body fat.

That millions of men, women, and children go hungry in one of the wealthiest countries in human history is a shameful reality. That this and other critical problems in America's food system get obscured from public view by bloated concerns over issues like obesity is all the more disgraceful.

8

Conclusion

The Perils of Perfectionism

*T*he best is the enemy of the good." That sentence kept popping into my mind as I listened to presentations at nutrition conferences and food-industry trade shows and spoke with a range of people about what they eat.

Attributed to Voltaire, in his *Philosophical Dictionary* of 1764, that aphorism neatly sums up what's wrong with much present-day thinking about food—from consumers who will buy only "authentic" ethnic foods or "bird-friendly" produce to scientists who enjoin us to eat only what they decree most healthful.

Call it the perfectibility trap, this impulse to idealize some foods while devaluing others that are plenty good for their intended purposes but don't further a pet view of proper eating.

Here's how crazy it has become: even as they struggle to feed 50 million needy Americans each year, hunger-relief organizations find themselves under attack for doling out "bad" fats, "fattening" carbs, and other supposedly suboptimal foods. These pressures to provide less pizza and more carrot sticks to the hungry come not primarily from the poor, but from well-to-do donors and volunteers at hunger agencies.[1]

"We have a name for them: nutritional imperialists," reports Robert Egger, founder of the D.C. Central Kitchen, an organization that distributes four thousand meals a day and trains homeless people for food-service jobs. While Egger welcomes any help he can get to obtain more donations of fresh produce, "that doesn't mean we should try to turn everybody into vegans overnight," he notes. "Take a look at the kids we serve. The Kitchen makes meals for dozens of after-school and weekend youth programs. We learned that if you give the kids a meal of healthy food—eggplant lasagna, salad, and an apple—they won't eat a thing. Kids will be kids. They'll take one look at the plate, pinch their noses in exaggerated disgust, and walk away hungry.

"But if you give them a slice of pizza—which we always have thanks to donations from local pizza chains—and you put the pizza next to a healthy salad, and a good piece of fruit and a cookie, they'll practically lick their plates clean. Once they see the pizza and cookie, they suddenly see everything else on the plate as edible."[2]

That Egger should have to apologize for serving pizza shows the folly of perfectionism. Even as nutritional imperialists condemn pizza, some of the world's most respected authorities on food commend it. "You remember when you were a kid and the lady held up the four basic food groups?" an eminent nutritional biologist at the University of California, the late Paul Saltman, liked to ask audiences. "Well, what the hell is a pizza? It's all of the above." Saltman declared pizza "the most nutrient-dense food in the supermarket, containing 44 nutrients." Jeffrey Steingarten, the prominent food critic, feels the same. "Pizza is a perfect food," he writes.[3]

By insisting that hunger-relief agencies distribute only foods they extol, nutritional imperialists continue a long tradition that dates back, in English-speaking countries, at least to the Irish potato famine of the nineteenth century. During the early years of the famine, the British did practically nothing to feed the several million starving Irish. Then, when the British finally did set up soup kitchens, they dished out sermons along with the broth. Protestant groups distributed tracts encouraging Irish Catholics to convert, and reformers preached alternatives to the Irish diet.

Alexis Soyer, one of nineteenth-century Europe's most prominent chefs, went so far as to publish a book, *Charitable Cookery*, in which he made clear that in assisting in the soup kitchens, he meant to do more than feed the hungry. He set out to correct their foodways by instructing them to replace their beloved potato—the perennial staple of the Irish diet and Irish national identity—with corn and fish, which he considered superior. Of twenty-two recipes in *Charitable Cookery*, just one included potatoes.[4]

Out of Left Field

I've fallen into the perfectibility trap myself. As someone who will eat (and love) anything from the chili dog at my favorite hot dog stand to everything on Thomas Keller's tasting menus, I don't succumb to the most common form of food perfectionism. I don't imagine that certain foods or cuisines are inherently superior to others. But as I discovered, it's easy to be a food snob anyway.

I realized this as I was interviewing Robert Forney, then president and CEO of America's Second Harvest. Headquartered in Chicago, this massive hunger-relief organization distributes 2 billion tons of food nationwide each year—an almost unimaginable quantity ("the equivalent of a convoy of trucks somewhere north of eight thousand miles long," Forney likes to say, "a good-sized mountain, bigger than most grocery-products companies"). America's Second Harvest also provides several million hot meals annually through a community-kitchens program and its "Kids Cafes" in more than one thousand schools and youth community centers.

Under Forney's leadership, America's Second Harvest fed more than 7 million Americans every day. So what was I doing criticizing him, midway through our interview, for accepting food from a particular grocery company?

Grateful for the 20 million pounds of food that Albertson's, the nation's second largest grocery store chain, donates each year, America's Second Harvest named it "Retailer of the Year" recently, ignoring, as I reminded Forney, the fact that the United Food and Commercial Workers Union strongly disagreed. In the late 1990s, the grocery workers' union filed suits in several states accusing Albertson's of coercing employees to work off the clock without pay and to refrain from filing workers' compensation claims. And for five months in 2003 and 2004, the very years Forney honored Albertson's, sixty thousand grocery store workers in Southern California went on strike against the company and two other chains over affordable health care coverage and a wage plan that paid newly hired workers less than existing employees.

Worn down and unable to support their families, the workers eventually caved in and accepted most of the grocery companies' conditions.[5]

"I'm not a labor expert, I'm a person who is responsible for finding food for Americans," Forney replied when I raised the issue. "Albertson's is one of the world's largest grocery retailers. I don't know how they deal with labor, but I know how committed they are to fighting hunger. And if they're on my side fighting hunger, I'm on their side."

The answer was slightly disingenuous. Forney was president and CEO of the Chicago Stock Exchange prior to taking over America's Second Harvest, and before that, of an information technologies firm, so he knows plenty about labor issues, and if he wants to look into a donor company's labor record, he certainly is in a position to do so. Furthermore, he had acknowledged earlier in our interview that low-paid workers routinely show up in food lines. Many of the people America's Second Harvest assists "have to make decisions between rent and food and medicine, or food and housing, or food and utilities," Forney said. "Those are the decisions that working poor people have to make, and unfortunately, that means that we're seeing a lot more people."

Each of the several leaders of hunger-relief agencies I consulted commented on the absurdity (immorality, some called it) of a wage system in which people who work forty or fifty hours a week cannot afford basic food and shelter for themselves and their children.

About 3 million Americans who work full-time have incomes below the poverty level. But does that shameful fact suggest that organizations like America's Second Harvest ought to refuse donations from companies that don't pay their employees a living wage? My own knee-jerk reaction is, *Hell yeah!* Organizations that aim to end hunger should not assist in face-saving publicity campaigns by companies whose business models perpetuate the problem.

On reflection, though, I recognize that this is just a left-liberal version of food perfectionism. To deny food to the hungry because it comes from a less than ideal company is as wrongheaded as nutritional imperialism.[6]

Political conservatives have an analogous and equally wrongheaded complaint. They criticize hunger programs for failing to discriminate between worthy and unworthy recipients. What good does it do, conservatives demand, to fill an addict's belly so he can go out and rob old ladies to pay for his next fix?

Both sides' concerns, while valid in their own right, become excuses for the inexcusable. To be sure, our society ought to protect its elderly and ensure its workers a living wage. But for practical as well as humane reasons, our top priority should be feeding the hungry. Hunger and malnutrition impair people's immune system, leaving them less able to hold jobs, more susceptible to illnesses they may transmit to others, and burdens on medical and emergency services. In children, food deprivations impair brain development, cause learning disabilities, and predispose kids to antisocial behaviors and unemployment later in life.[7]

A nation that has the resources to feed its hungry has a moral obligation to do so, and those who help fulfill that obligation ought to be praised, not condemned because some of the foods they hand out, or the providers or recipients of those foods, do not conform to someone else's ideals.

Whose Guidelines Are These?

That's the unfortunate thing about dietary idealists. They impose their standards on others, in particular, poor and working people. Food reformers, whether liberal or conservative, typically go after the foods served in hunger programs, fast-food chains, and low-income homes, not in country clubs, upscale restaurants, and Park Avenue penthouses.

Sometimes reformers merely preach or scold; at other times they literally *legislate* what poor people can eat, by way of government mandates like the Department of Agriculture's *Dietary*

Guidelines for Americans. For most of us, that title is accurate; we can choose to be guided by the government's directives for how much fat, salt, sugar, cholesterol, fruit, dairy, and grains to include in our diets, or we can ignore them. But many underprivileged Americans don't have that option. Programs they rely upon for food—school meal programs; government-run hospitals; and the Women, Infants, and Children Supplemental Nutrition Program—are required to comply with the *Dietary Guidelines.*

Such mandates would never become law if policy makers themselves had to adhere to them, as a reporter for the *Los Angeles Times,* Roy Rivenburg, learned when he tried to stick to the *Dietary Guidelines* for just ten days. Even though Rivenburg had the luxury of fortifying himself in advance on pancakes, hoagies, ice cream, and in general "enough sugar, butter, salt and fat to clog the arteries of every man, woman, child and dog in Los Angeles," he found the regime insufferable.

Once on the diet, Rivenburg recalls, "it didn't take long to run afoul of the guidelines. But I was shocked it happened at breakfast, which I considered my fail-safe meal." His customary morning menu—cereal topped with dried cranberries; a banana, juice, English muffin with jam, and chai latte—he had always considered healthy. But the *Dietary Guidelines* say not. "The muffin, despite a label that said 'Healthy Multi-Grain,' wasn't made from whole grains, which are more nutritious than their processed cousins. And before I could sip any chai, I discovered the cereal, jam and cranberries busted my added-sugar quota for the entire day," Rivenburg reports.[8]

That quota and the *Guidelines'* other commands ostensibly come from science. A group of Ph.D.'s and M.D.'s from places like Johns Hopkins, Columbia, and UCLA drafted the documents, which they peppered with phrases like "science-based food guidance," "science clearly shows . . . ," and "the latest science." But in reality, the *Guidelines* are, as Marion Nestle, professor of nutrition, food studies, and public health at New York University, maintains, "best understood as a committee report, the result of the interplay of give-and-take, bullying, boredom, and eventually compromise among a group of people who entered the process with differing opinions and agendas."

While serving on a *Guidelines* advisory committee in the mid-1990s, Nestle experienced firsthand the intense lobbying by powerful sectors of the food and dietary-supplement industries, members of Congress, and assorted advocacy organizations. Groups as diverse as the National Dairy Council and the Congressional Black Caucus showered conflicting scientific evidence on Nestle and her fellow committee members. No sooner would the milk producers present studies showing their product to be a nearly perfect food than the Black Caucus would offer evidence that most people of color are lactose-intolerant after childhood.

Muddying matters all the more, several members of the *Guidelines* advisory committee had ties to the dairy, egg, and meat industries, Nestle discovered; and as I later learned, some of the advocacy groups concealed their ulterior motives as well. The Physicians Committee for Responsible Medicine (PCRM), which helped organize the Black Caucus's antimilk campaign, for example, was hardly a disinterested evaluator of the scientific literature on dairy products. When testifying before government committees, the PCRM cites numerous studies in support of its claims, but carefully avoids revealing that in reality, it is an organization of animal rights activists whose ultimate goal is to eliminate meat, dairy, and fish from the American diet.[9]

Industry groups don disguises as well. A large body of evidence shows that healthy people get little benefit from the vitamin and herbal supplements they take in hopes of bettering their diets. Tens of millions of Americans have experienced adverse reactions to nutritional supplements, and some supplements, including seemingly innocuous ones like vitamin A and vitamin E, can cause serious complications and deaths in high doses. So one might wonder why a group called the Council for Responsible Nutrition (CRN) would urge the *Dietary Guidelines* committee to recommend nutritional supplements to the general public.[10]

But the CRN is as much about responsible nutrition as the GreenFacts Foundation, funded by chemical companies, is about responsible environmental policy. Owned and operated by the $20 billion supplement industry, the CRN exists to persuade policy makers and reporters that supplements are safe and salutary, indeed superior to actual food. "In a society where the majority of us are eating too much, too often, there is a potential danger in ignoring the fact that conventional foods, because they inevitably provide calories as well as nutrients, may not always be the optimal mechanism for increasing the intake of specific nutrients," the president of the CRN said with a straight face at a hearing on the 2005 Food Guidance System (the graphic popularly known as the "food pyramid").

Big Claims in Small Packages

The promise of dietary perfection by means of a pill, central to the marketing strategy of the supplement industry, capitalizes on an age-old belief in what prominent food historian Warren Belasco calls "minute foods with enormous powers." Like many peoples before us, we Americans imagine it possible "to distill vital essences into a highly condensed 'elixir of life,'" Belasco has said.[11]

Throughout the nineteenth and twentieth centuries, numerous writers predicted that scientists would soon create what the author of an article in *Popular Science Monthly* in 1936 described as "food pills that contain everything necessary for life." A less futuristic version of this fantasy, common throughout human history, attributes magical properties to small foods. Cultures around the globe have credited grains of corn, wheat, or rice with extraordinary powers that belie their size. Beans, nuts, and grapes have enjoyed special standing as well. Each has been thought to contain the essence of life or the key to strength and prosperity.

In twenty-first-century America, the vitamin industry is far from alone in exploiting this way of thinking to sell goods. We saw earlier how makers of functional foods portray their products as exceptional on account of the oat bran or soy isoflavones they contain. Fruits and nuts are marketed the same way. Olives and their celebrated oil, for instance, "quickly satisfy hunger, lead to fewer total calories ingested at mealtime . . . and decrease rates of cardiovascular disease and cancer," according to the trade association for the California olive industry.[12]

Larger and lower-status fruits get sold this way as well. "Practically a multivitamin unto itself," proclaims the National Watermelon Promotion Board about its product in a brightly colored brochure a young woman handed me at a trade show for the fresh produce industry. Loaded with vitamins A, B_6, and C, watermelons bolster brain function, strengthen bones, and help prevent cataracts, colds, and bad moods, the brochure asserts.

Not only that, "watermelon is the lycopene leader among fresh fruit and vegetables." Taking direct aim at tomatoes, whose marketers play up their fruit's high lycopene content and cite studies showing that this antioxidant prevents cancer and heart disease, the Watermelon Board made known that a single serving of watermelon contains as much lycopene as four and a half tomatoes.

What a lovely irony: a fruit long associated in the U.S. with lower-caste African Americans gets recast in its marketing materials as an ideal food for white folks. (Everyone pictured in the brochure appears to be white.)

How the Almond Got Its Halo

Anything edible can plausibly be christened an ideal food. Having amassed hundreds of promotional pieces akin to the Watermelon Board's brochure and reviewed at least as many studies that food marketers and dietary reformers cite in support of their claims, I can say with assurance that credible evidence exists to glorify even the most humble of foods.[13]

To make an ordinary food seem extraordinary, you don't need to change the food itself; you just need to change how it is viewed, a feat that food marketers pull off by culling auspicious findings from studies, or testimony from respected authorities, and packaging it in persuasive ways. If their food contains something that conflicts with a prevailing dogma, be it fats, sugars, carbs, calories, or whatever else, they talk instead about how it provides the "correct types and balance of essential fats," as an ad for walnuts puts it, or "minerals necessary for proper water balance inside the cells of the body," as the watermelon brochure boasts.

Or they offer proof that the product complies with the gospel of naught despite its reputation. The Hass Avocado Board, aware, no doubt, that many people correctly perceive their product as high in fat and calories, counters throughout its promotional materials, "Hass avocados are cholesterol-free, sodium-free, and low in saturated fat."

Or a marketer will exploit the obesity panic, even if the food it's hyping is generally considered fattening. The U.S. Potato Board, which represents ten thousand potato growers and manufacturers, has had a campaign for several years to distribute copies of "The Truth about Carbs," a brochure published by Weight Watchers that argues for "the important role carbs play in healthy weight loss." "Give them," the Potato Board exhorts its members about the brochures, "to key retail accounts to pass out at new store openings . . . include them in your business mailings . . . involve community organizations like Rotary and PTA."

The National Peanut Board, ignoring the high calorie and fat content of peanuts and their popular by-product, peanut butter, goes on for eight pages in the *Peanut Press*, a glossy promotional piece designed to look like a newsletter, about how its product is just what the doctor ordered for anyone who wants to lose weight. "Just a handful of peanuts or a tablespoon of peanut butter can be the kind of fuel you need to go the distance," the Peanut Board contends.

Its proof? "Have you ever watched those highly conditioned athletes enduring the rigors of the Ironman Triathlon?" the *Peanut Press* asks. "Look closely, and you're likely to see many of them fueling up with peanuts and peanut butter."

No matter how closely I watched the Ironman event on ESPN, I couldn't spot a single peanut. Presumably the contestants fueled up off camera. But in mentioning triathletes, the Peanut Board makes use of another common ploy of food marketers: elevating the social standing of a product by associating it with well-regarded people. Similarly, at one of the food-industry trade shows I attended, marketers for a competing nut linked their product to another group that's commonly called upon to lend cachet—celebrated chefs. The Almond Board of California handed out, along with a fifteen-page treatise titled "Health Aspects of Almonds," a slice of almond-and-polenta cake drenched in almond liqueur and capped with crème fraîche. The Almond Board provided, too, recipe cards developed by well-known chefs. The cake recipe, it turned out, came courtesy of the Food Network's hyperkinetic "naked chef," Jamie Oliver.

The Almond Board makes expert use as well of another tried-and-true method for positioning a food apart from its similarly hyperbolized competitors. It depicts almonds as great not just in their own right, but for bettering other foods, as a way to make treyf safe. In an ad in a trade magazine for food developers, below the headline, "THANK YOU, SCIENCE," and above pictures of fruit drinks and cereals topped with almond slices, the Almond Board reveals that "78% of consumers recently agreed that a product containing almonds is better nutritionally. Which means that now, the added value of almonds goes even further than crunch, flavor and appeal—they can give any product a healthy halo."[14]

An Orgy of Evidence

That brings up a question I have addressed at several points in this book: Why *not* rely on science for advice about what to eat? Although food marketers and advocacy groups shamelessly hijack science for their own opportunistic ends, this does not mean that scientists cannot say which foods are best.

It's hard to deny the wisdom of relying upon nutritional scientists for *some* kinds of dietary guidance. If you don't want scurvy, you'd better consume some vitamin C. If you're planning to become pregnant, you need enough folic acid to protect your child against neural tube defects. If you're suffering from iron-deficiency anemia, you are well advised to eat raisins, beans, liver, eggs, and other foods high in iron.

But deficiency disorders are very different from obesity, heart disease, cancer, and other chronic diseases that Walter Willett and friends say we can thwart by eating their favored foods. Chronic diseases are caused, as we have seen, not by a missing nutrient, but by a complex interplay of genetics, stress, physical inactivity, and a host of other factors. Undoubtedly, diet plays a role, but science is ill-equipped to tell us how much of a role, or which ballyhooed foods are ultimately the most healthful. Depending upon which aspects of its chemical composition one focuses upon, any nutrient-rich food can come off as magical or deficient. Beef is a good example. Viewed one way, it is loaded with essential vitamins, minerals, and protein, a fact that the beef industry presents frequently in ads to consumers and testimony before government nutrition panels. Viewed another way, beef lacks fiber and has an excess of "bad" fats and calories, as groups like the Physicians Committee for Responsible Medicine routinely point out.

In planning meals to prevent disease, "think broccoli, tomatoes, blueberries, sweet potatoes, and garlic," the PCRM urges, and notwithstanding the body odor and frequent trips to restrooms such a diet would produce, heaps of studies support the ultra–low-fat, high-fiber view.

But was a spokesman for Atkins Nutritionals off base when he testified at a government hearing on the food pyramid that "much as the low-fat advocates and the animal rights activists would like to believe otherwise, the accumulating body of scientific evidence can no longer be ignored"? He and his colleagues cite an enormity of studies that corroborate their pro-protein, anticarb theory. When I interviewed the vice presidents for product development and marketing at the Atkins organization, they used the word "science" almost as often as my college undergraduates use "like" and "whatever," and they referred me to bibliographies longer than an eighteen-wheeler.[15]

As I write this, the Atkins organization is funding more research into the glories of low-carb living, and both the Wheat Foods Council and the PCRM are sponsoring opposing studies. The bread industry group has increased its financial support to scientists who study the health benefits of grain in the American diet. The PCRM, meanwhile, has set up a Web site—www.atkinsdietalert.org—to collect information from people who believe that Atkins-style diets have made them ill.[16]

How Much Is That Scientist in the Window?

In principle, these self-interested efforts can be offset by research from independent scientists with open minds and ties to neither food companies nor advocacy groups. But good luck finding such folks. Time and again, scientists I had assumed were free agents turned out to be otherwise.

A case in point: following the release of the 2005 revision of the *Dietary Guidelines,* numerous news stories included a pair of scientists lauding the government's recommendation to eat more grains. The two scientists depicted whole grains as practically a prophylactic against heart disease, cancer, and strokes. Both men have impressive credentials. David Kessler, a former director of the Food and Drug Administration, is dean of the University of California, San Francisco, Medical School; and Meir Stampfer is professor of epidemiology and nutrition at Harvard.

Watching the news coverage, I had no idea that these guys were pimping for Big Food. On the contrary, while I knew that Kessler had received six-figure speaking fees from drug companies in his post-FDA days, he never struck me as someone involved with the food industry. Indeed, over the years he has been highly critical of some of what it sells. And as for Stampfer, I was under the impression that ever since an exposé in the 1970s, Stampfer and his colleagues in Harvard's nutrition department had become wary of flacking for the food industry. The exposé, written by Michael Jacobson of the CSPI and two other muckrakers, had depicted the Harvard department as "riddled with corporate influence."[17]

Jacobson's report was particularly critical of the chair of the nutrition department at the time, for having testified to a congressional committee investigating the nutritional content of cereals that "breakfast cereals are good foods," even as his department was receiving hundreds of thousands of dollars from Kellogg's and Nabisco.

I learned of Kessler's and Stampfer's ties to the present-day cereal industry by chance. In the course of an unrelated Google search, I happened on the source of their widely quoted comments—a video news release (VNR) prepared by General Mills as part of a multimillion-dollar advertising and public relations campaign proclaiming the company's Cheerios and Trix cereals just what the *Guidelines* ordered.

Indistinguishable from what usually appears on newscasts, VNRs are ready-to-air news stories that companies or their PR firms produce and send to TV news outlets. The one prepared by General Mills began with video of supermarket shoppers examining various fruits, meats, and cereal boxes, and a voice-over narration by a woman who sounded like every other TV news reporter. "The federal government has changed its dietary guidelines," she said, and she gave a quick summary of the recommendations before turning to Kessler, who held forth on the wonders of whole grain and singled out General Mills as being in the forefront of efforts to bring those little elixirs of life to America's breakfast tables.

The two-and-a-half-minute VNR concluded with Stampfer in his office at Harvard, sounding like a disinterested scientist. "By increasing consumption of whole-grain foods from the typical one serving a day to three or more, our research shows that we can expect a substantial reduction in the number of premature deaths in America," he said. Stampfer went on to accord grains a place in a holy trinity: "Like fruits and vegetables, whole-grain foods offer a complete package of naturally occurring vitamins, phyto-nutrients, minerals, fiber, and protein. All of these are key components of a healthy diet that can help prevent disease."

"I'm Danielle Addair," the pretend reporter signed off at the end of the fake news report, which TV stations on tight budgets could include in their newscasts as if they had put together the story themselves. Larger stations and the national networks, meanwhile, could extract Kessler's or Stampfer's sound bites and insert them in their own reports.

Precisely how many TV news outlets ran all or part of General Mills' VNR, I do not know, but presumably several dozen at least. Around this same time, when the Government Accountability Office investigated the airing of a VNR prepared by the White House in support of a Medicare bill, the GAO found it had aired at least fifty-three times at forty stations throughout the country. According to several journalism watchdog groups and public relations executives, most of the nation's TV news operations regularly used VNRs during this period.[18]

The Terroir of Genetically Modified Foods

But let's give the scientific quest for an ideal diet the benefit of the doubt. Suppose the biases and conflicts of interest I have cataloged are somehow eliminated in the future and genuinely objective scientists take up the quest: women and men with no stake in the beef-versus-broccoli debate, no ties to cereal makers or the anti-industry advocacy groups that assail them; indeed, no soft spot for particular foods. Imagine, too, that these scientists have found their way around the myriad methodological problems involved in getting accurate information about what people eat and sorting out the physical effects of a meal from the psychological and social effects.

Surely *then* we could rely upon scientists to tell us which foods are best, couldn't we?

We probably could, but here's the kicker. That we *could* does not mean we *should*. Were scientists to determine reliably that a diet I find dull will extend my life by a couple of years, should I rely on their criteria for judging the quality of a meal over those of the Ruth Reichls, Jeffrey Steingartens, Frank Brunis, Jonathan Golds, and other great food critics?

Or if the futurists turn out to have been clairvoyant, and meals become available in the form of pills, should I choose them over conventional foods?

Findings from science, regardless of how consistent or credible, cannot settle such value-laden questions. Anyone who would imagine otherwise need only consider the European reception for American-grown soy. In Europe as in the U.S., soy routinely appears on lists of optimal edibles by virtue of its high-quality protein, unsaturated fat, antioxidant properties, and purported favorable effects on cholesterol and symptoms of menopause. For health-focused Americans those are more than enough reasons to seek out soy milk, soy oil, soy burgers, soy breads, soy everything.

Soy plantings occupy 75 million acres of U.S. farmland. So why do Europeans spurn the stuff? The reason is not that they are more aware than we that excessive consumption of soy may be damaging. (They're not.) Europeans do not reject soy products in general. They reject U.S. soy because it fails to satisfy a criterion that their shared experiences and cultural values have led them to consider crucial—the absence of genetic modification.

Americans blithely eat genetically modified (GM) foods. With 80 percent of the soy and 40 percent of the corn grown in the U.S. genetically modified, and these staples finding their way into everything from mayonnaise to fruit drinks, three-quarters of the processed products in our supermarkets contain GM ingredients. We insist on all sorts of other *nons* (nonfat milk, nonsugar sweeteners), but so far, GM has not been among the *nons* that large numbers of Americans demand.

Across the Atlantic, by contrast, anti-GM sentiment is so widespread that politicians win elections by taking a stronger stand against "Frankenfoods" than their opponents. In the late 1990s and early 2000s, the European Union banned GM foods outright. It partially lifted the ban in 2004, but with strict rules that foods containing even a tiny amount of genetically modified material be clearly labeled. In surveys and interviews, large numbers of Britons, French, Italians, and Austrians speak of GM as an abomination against nature and tradition, and obviously unhealthy.[19]

They appear to be unswayed by the many knowledgeable scientists in Europe and elsewhere who have presented evidence to the contrary. Declared "entirely safe" by the U.S. Food and Drug Administration, some GM foods are arguably better than their conventional counterparts. Faster-growing and pest- and disease-resistant, these crops require less use of pesticides and less tilling of the soil, thereby decreasing soil erosion, water pollution, and, possibly, greenhouse emissions. What's more, by means of genetic modification, a food's best features, like its flavor or vitamin content, can be increased, while its undesirable attributes, like the ability to cause allergic reactions, can be reduced.

"Whatever fears I might have of possible allergic reactions to food produced from genetically modified organisms," biologist Richard Lewontin of Harvard has remarked, "they are not more unsettling than the allergies induced in me by the quality of the arguments about them." Noting that conventional plant breeding, which has gone on for centuries, sometimes produces foods that make people ill, while as far as we know, genetically engineered foods have harmed no one, Lewontin suggests that those who oppose GM foods succumb to "a false nostalgia for an idyllic life never experienced."[20]

The nostalgia is more pronounced in Europe than in the U.S. partly for geographic reasons. As journalist Kathleen Hart noted in a book on the GM controversy, "Many Europeans, even city dwellers, have close ties to friends and relatives in the countryside. Throughout Europe patches of farmland are never distant from cities and towns, whereas many Americans are far removed from the vast swaths of farming regions of the country where most of the nation's food is grown."[21]

And Europeans shared a frightening experience in the mid-1990s that we Americans only heard about—an outbreak of mad cow disease in Britain that spread to other European countries. For more than a decade after the disease was first diagnosed in cattle in 1985, numerous public officials, scientists, and government commissions described the risk to humans as almost nil. Little wonder, in the decade following the British government's admission in 1996 that the disease had spread to humans, that many in Britain and the rest of Europe would be wary of assurances about GM food. In surveys during this period, Britons cited mad cow disease as the principal reason for their skepticism about GM food.[22]

There is another cultural factor as well behind Europeans' disdain for GM. Many in Europe believe the quality of a food results largely from where it hails. This notion is captured by the French word *terroir*, which Slow Food's founder Carlo Petrini defines as "the combination of natural factors (soil, water, slope, height above sea level, vegetation, microclimate) and human ones (tradition and practice of cultivation) that gives a unique character to each small agricultural locality and the food grown, raised, made, and cooked there." Though most familiar to wine enthusiasts who recognize characteristic differences between, say, Burgundy and Bordeaux, the concept applies to foods as well, as aficionados of artisanal cheeses from diverse regions of Europe can attest.[23]

But what is the *terroir* of GM foods? In the eyes of many Europeans, it could not be worse. That GM originated in America, the birthplace of fast food and the nation M. F. K. Fisher called "taste-blind," is bad enough. Worse, genetically modified foods do not even come from farms or traditional food companies. They come from corporations like Monsanto, the U.S. chemical company that brought the world synthetic pesticides, saccharin, and PCBs.

Family Magic

We Americans romanticize in our own way. We're saps for phrases that begin with "family." Attach that word to a noun, and poof! it becomes charmed, as witness the mileage Republicans have gotten from "family values" over the past quarter century, food companies get from "family farms," and journalists and advocacy groups get from "family meals."

In consumer surveys, substantial numbers of Americans say they consider the products of family farms superior to those from commercial operations. Aware of that, food makers proclaim their support of family farmers in advertisements and adorn their product wrappers with labels such as "Family Farmer Cheese."[24]

The label hardly provides a reliable metric of quality or distinctiveness, however. If gauging foods by what they do or do not contain (GM, particular nutrients) has drawbacks, betting on the family farm is even more dicey. "The distinction between corporate and family-owned farms does not hold much water," notes Julie Guthman, a professor of community studies at the University of California who spent several years studying organic farms throughout California.[25]

No one denies that small farming operations—the places that provide much of the tastiest produce at farmers' markets—deserve support, but Guthman's study and others make plain the naïveté of prejudging the goodness of a farm or its products based on whether it is family owned. While some family farms bear a resemblance to the agrarian ideal in the ads—noble yeomen lovingly tending their twenty-acre plot—many are exponentially larger, dirtier, and more mechanized. Some of the nation's largest farming operations are family-owned partnerships, Guthman discovered, and many family-owned farms have created closely held corporations to gain special tax and legal advantages.[26]

Most members of farm families these days do little or no farming. Nationwide, 90 percent of farm households' income comes from nonfarm jobs, and on many so-called family farms, at most one member of the family devotes his or her time to farming. The rest of the family works off the farm. In some cases, *no one* from the family does farmwork full-time. The place technically qualifies as a family farm because it is family-owned, but farming is at most a hobby for everyone in the family. Hired hands till the soil and tend the books.[27]

Something else we Americans imagine turns to gold with the addition of the word "family": meals. "Experts are finding that making family meals a priority is more than worth the effort," reports Mimi Knight in the magazine *Christianity Today*. Bemoaning what she describes as a sea change over the last half century, from families dining together routinely to the present situation in which parents are so busy with work, and children with after-school activities, that no one has time for family meals, Knight advises her readers not to make that mistake in their own homes. "Consider," she writes, "a recent survey from the National Merit Scholarship Corporation. The NMSC profiled National Merit scholars from the past 20 years trying to find out what these stellar students had in common. They were surprised to find that, without exception, these kids came from families who ate together three or more nights a week."[28]

Writers for many other magazines and newspapers have made similar claims in recent years. They hold out the family meal as a magic potion for ills of all sorts, from poor school performance and the example I discussed in the previous chapter (consumption of fast food) to eating disorders and parental divorce. A writer in the alternative magazine *Utne Reader* depicted family dinners as "the cornerstones of human civilization," more important to children's success and well-being than household income, parents' education, or neighborhood affluence.[29]

He, too, cited the National Merit study—a research endeavor of truly mythical proportions. Over the years, I have encountered more than a few studies so poorly conceived that they should never have existed, but the NMSC study is unique. It *re-*

ally never existed. Having come upon numerous summaries of the study, I figured I should read the original, so I wrote to the NMSC for a copy.

Here is their reply:

Dear Prof. Glassner:

We have also seen references of this study in various articles over the last 10 to 12 years. However, we have no idea of the origin of this study. All I can tell you is that it did not originate from National Merit Scholarship Corporation. No such study was ever done here. I'm sorry I cannot help you further. Good luck in locating the source.

Sincerely,

Eileen Artemakis
Dept. of Public Information
National Merit Scholarship Corporation

Not only is the NMSC study a myth, so is the alleged demise of family meals. Mourners can take cheer; present-day pronouncements of the death of the family meal are only the most recent of many over the last two centuries. In 1838, Sarah Ellis, a chronicler of family life in Victorian England, reported that wives and children so seldom saw their husbands and fathers, "we almost fail to recognize the man." In the 1920s, an American sociologist quoted a father lamenting, "It's getting so that a fellow has to make a date with his family to see them."[30]

Today, the family dinner is no more a thing of the past than it was when those writers made their claims. Three out of four American families with children under eighteen eat dinner together at least five nights a week, roughly the same number as in the 1950s.[31]

Whatever the cause of America's social problems, it is not a cutback in the consumption of family meals; but this reality hasn't prevented advocacy groups and journalists from bowing to every study, mythic or actual, that enthrones the family table. In 2004, they touted a university-based study ostensibly showing that adolescent girls who take meals regularly with their families are less likely to exhibit symptoms of eating disorders. Read the fine print in that study, though, and you find that the researchers themselves raise the chicken-and-egg problem. They call it "highly probable" that teens with eating disorders are less likely to join their families for dinner, and they observe that simply dining together is not enough to help girls avert eating disorders. The meals need to take place "in an enjoyable atmosphere that is free from conflict around food or other issues," the researchers report. Indeed, as they point out, other studies have found that unharmonious family meals *contribute* to the development of eating disorders.[32]

Just as no single food is ideal for everyone, neither is any particular grouping at mealtimes. Those who idealize the family meal close their eyes to the reality that many children and adults would be better off missing the tensions that exist in their households at mealtime. The French historian Luce Giard may go too far in describing family meals as "fierce power struggles," but anyone who saw the 1999 movie *American Beauty* probably recalls the dining room scene in which the uptight wife, sullen adolescent daughter, and funereal atmosphere of the meal drive the man of the house, played by Kevin Spacey, so mad he flings his dinner plate at the wall.

If dinners that dreadful are uncommon, so too are suppers as charmed as promoters of the family meal would have us believe. As Darra Goldstein of Williams College, a leading food scholar, notes, "The home-cooked meal has come to stand for many values we now fear have vanished. Yet rarely outside of a Norman Rockwell painting was that idealized family meal the perfect gathering it's now made out to be."[33]

Eat and Let Eat

Why do we deify some meals and some foods, and demonize others?

That question has guided much of my discussion throughout the book, and I have already offered some answers. Having lost faith in medicine, we turn to things we can change, like our diet, and invest them with the power to make us ill or well. And having lost faith in what the big food companies and restaurant chains dish out, we christen alternative fare "natural" or "authentic" and revere it beyond its due.

As Goldstein's observation suggests, we long for safe havens in an unsettled world. In forcing our teen to the family table, or filling our plate with foods consecrated by nutrition professors at Harvard, we hope to escape not just diseases and weight problems, but a more sweeping pathology as well—what Steven Shapin, a historian of science, refers to as "the bad order of society." Shapin points out that programs as seemingly dissimilar as the Atkins diet, vegetarianism, and the organic and Slow Food movements all preach a common message. "A bad society makes bad food, and bad food makes badly motivated and badly functioning people," he says, summing up their shared dogma.

The dietary-reform movements obviously differ in which foods they deem best and worst, but each portrays its preferred diet—meatless, carbless, artisanal, organic, or what have you— as the route out of the bad order of society and a safe haven from the blundering masses who consume the bad foods that society spews forth.[34]

Dietary reformers capitalize on our tendency to take as literal truth the old adage "You are what you eat," that what and where someone eats determine the kind of person he or she is. In an experiment at Arizona State University, psychologists showed students a set of photographs of people their age. To some of the students, they described the people in the photos as eating "good" foods like fruits, wheat bread, and chicken. Other students were told that the people in the pictures ate doughnuts, hamburgers, French fries, and fudge sundaes. Though the students were shown identical photographs, they ranked the people in them as less attractive, likable, and moral if they were told they ate the "bad" foods.[35]

In the larger world outside psychologists' labs, these sorts of comparisons extend beyond specific individuals to entire categories of people, and they cut two ways. We judge groups by what they eat, and we assess the quality of a food partly by who eats it. The English demeaned the Irish as "spuds" and "potato heads," and the potato's association with the Irish and other low-status groups helped make it a low-status food. In parts of the U.S., racist whites refer to Hispanics as "bean burritos," and the social standing of "Tex-Mex" cuisine is commensurately low.[36]

Tellingly, as a group's social status improves, so does that of its food. Today, top-rated restaurants in America's major cities proudly serve Italian salamis and sausages, and the Slow Food organization holds tastings and seminars devoted to them. But in the early twentieth century, when prejudice against Italian immigrants ran high, respectable chefs would never serve such meats, and prominent reform groups and nutritional organizations circulated pamphlets warning that they harmed the stomach and provided no nourishment.[37]

Dietary reformers of that period went after a long list of foods beloved by immigrants from Southern and Eastern Europe. "Nutritional science told them," historian Harvey Levenstein reports of American do-gooders back then, "that the essence of European economical cooking—the *minestras* and *pasta-fagioles* of Italy, the *borschts, gulyashen,* and *cholents* of Eastern Europe—were uneconomical because they were mixtures of foods and therefore required uneconomical expenditures of energy to digest. Strong seasonings that made bland but cheap food tasty were denounced for overworking the digestive process and stimulating cravings for alcohol.

"Nutritional science reinforced what their palates and stomachs already told them: that any cuisine as coarse, overspiced, 'garlicky,' and indelicate-looking as the food of central, eastern, and southern Europe must be unhealthy as well."

Prejudices dressed up as science are still prejudices. As Levenstein notes of the reformers of a century ago, "To most of the native-born daughters of the middle and upper class, who preferred their own food awash in a sea of bland white sauce, and for whom 'dainty' was the greatest compliment one could bestow on food, one whiff of the pungent air in the tenements or a glance into the stew pots was enough to confirm that the contents must wreak havoc on the human digestive system."[38]

Were dainty dinners truly better than the chopped liver with onions, roast chicken with prunes and sweet potato pie, and raisin-laced noodle kugel my grandmother served around that time? Anyone who thinks so should, as her husband, my grandfather, used to say, have his head examined. So should people who convince themselves that the foods revered by today's dietary reformers are superior to beloved dishes at immigrant eateries of our day. Baked fish in a fat-free sauce with broccoli on the side can be quite pleasing if you're in the mood for something light and the fish is fresh and flavorful. Just don't try to tell a Tex-Mex devotee that it beats a great carne asada burrito with refried beans, or a fan of Vietnamese food that he or she should give up *bò 7 món,* or seven courses of beef.

My favorite meal at Vietnamese restaurants, *bò 7 món* begins with a salad of marinated beef, concludes with a beef porridge, and includes along the way sausages, fondue, and courses of grilled and steamed beef served on skewers, as meatballs with peanuts, onions, and lemongrass, and wrapped in an aromatic green called *la lot.*

Personally, merely thinking about that feast improves my well-being.

Acknowledgments

Above all, I thank my wife, Betsy Amster, for her love, support, and extraordinary advice on many drafts of each chapter over the five years I worked on this book. I am also deeply grateful to my agent and friend, Geri Thoma; Daniel Halpern and Emily Takoudes, insightful editors at Ecco; the brilliant chefs, food critics, historians, investigative journalists, and scientists I cite throughout the book; students in my seminar, "Food and Society," and graduate assistants Natali del Conte and Pamela Leong, who helped with library research; and my boss, USC's provost, C. L. Max Nikias, for granting me the time to finish the manuscript.

Ordinarily at this point, I would acknowledge as well a longer list of friends and colleagues for their suggestions and encouragement. This time I have chosen to thank them privately, however, because in my mind, one person overshadows everyone else. David Shaw, a Pulitzer Prize–winning writer for the *Los Angeles Times* and a treasured friend whose passion for food and life exceeded that of anyone I have known, died of a brain tumor just as I was completing the manuscript.

On the eve of David's memorial service, his wife, Lucy Stille, threw a small dinner party at their home. The menu featured David's favorite "down home" foods from around Los Angeles: pastrami sandwiches from Langer's Deli; sausages from the Wiener Factory; fried chicken from Roscoe's House of Chicken and Waffles; and the dishes Lucy asked me to pick up for the party: ribs and links from Phillips Barbecue. As I waited for the order, I recalled the many times David and I would pick up lunch there, go to the park a few blocks away, and talk.

At the memorial service, Jonah Lehrer, a young writer and family friend, said what I suspect many of us in the room were thinking: "David knew that we have to eat, and so what we eat may as well make us happy. But he also knew that the happiness we find in the breaking of bread and the drinking of wine is not in the bread or in the wine but in whom we share it with. David taught me how to eat, and every time I eat like him, every time I look around and feel like I am celebrating something, I will think of him and I will miss him."

Notes

Preface: Eating Is Believing

1. Warren St. John, "Cancer? Suicide? Politics? That's Hilarious!" *New York Times*, December 12, 2004.

2. The surveys are from the International Food Council, Functional Foods Attitudinal Research, with summary data posted at the council's Web site, www.ific.org.

3. Greg Winter, "Contaminated Food Makes Millions Ill Despite Advances," *New York Times*, March 18, 2001.

4. Leslie Brenner, *American Appetite* (New York: HarperCollins, 1999).

5. See Howard Schutz and Oscar Loren, "Consumer Preferences for Vegetables Grown Under 'Commercial' and 'Organic' Conditions," *Journal of Food Science* 41 (1976): 70–73; Patrick Stewart, "Consumption Choices Concerning the Genetically Engineered, Organically Grown, and Traditionally Grown Foods," *Knowledge, Technology & Policy* (2000): 58–72; J. M. Darley and R. Fazio, "Expectancy Confirmation Processes Arising in the Social Interaction Sequence," *American Psychologist* 35 (1980): 867–81; Jana Lindsay et al., "Fat Content Labels Influence Acceptability Ratings," *Journal of Family and Consumer Sciences* 93 (2001): 54–55; Debra Zellner et al., "Effect of Temperature and Expectations on Liking for Beverages," *Physiology & Behavior* 44 (1988): 61–68; Hendrick Schifferstein and A. Kole, "Asymmetry in the Disconfirmation of Expectations for Natural Yogurt," *Appetite* 32 (1999): 307–29.

Chapter 1: False Prophets

1. Guidelines, Swedish and Thai study, and related research: *Tufts University Health and Nutrition Letter*, October 2000. The original study: Leif Hallberg, E.

Bjorn-Rasmussen, et al., "Iron Absorption from Southeast Asian Diets. II. Role of Various Factors That Might Explain Low Absorption," *American Journal of Clinical Nutrition* 30 (1977): 539–48. See also Paul Rozin, Claude Fischler, et al., "Attitudes to Food and the Role of Food in Life in the USA, Japan, Flemish Belgium, and France," *Appetite* 33 (1999): 163–80.

2. Marjaana Lindeman and K. Stark, "Loss of Pleasure, Ideological Food Choice Reasons, and Eating Pathology," *Appetite* 35 (2000): 263–68.

3. Molly O'Neill, "Can Foie Gras Aid the Heart?" *New York Times,* November 17, 1991; Angela Smyth, "Do Not Beware the Fattened Goose," *Independent* (London), December 24, 1991.

4. Rozin et al., "Attitudes . . ."; Laura Fraser, "The French Paradox," *Salon,* February 4, 2000. For alternative (including diet-based) hypotheses, see Malcolm Law and Nicholas Wald, "Why Heart Disease Mortality Is Low in France," *British Medical Journal* 318 (1999): 1471–80, and in the same issue, a commentary on that paper by Meir Stampfer and Eric Rimmin.

5. "Periscope," *Newsweek,* September 11, 2000, p. 6.

6. Walter Willett, *Eat, Drink, and Be Healthy* (New York: Simon & Schuster, 2001), p. 54.

7. Warren Belasco, "Future Notes: The Meal-in-a-Pill," *Food and Foodways* 8 (2000): 253–71 (quotes appear on pp. 260 and 261).

8. Willett, *Eat, Drink, and Be Healthy,* pp. 19–20.

9. Redcliffe Salaman, *The History and Social Influence of the Potato* (Cambridge, UK: Cambridge University Press, 1949); Ellen Messer, "Three Centuries of Changing European Tastes for the Potato," pp. 101–13 in Helen Macbeth, ed., *Food Preferences and Taste* (New York: Berghahn, 1997). Nicholson quote is from Barbara Haber, *From Hardtack to Home Fries* (New York: Free Press, 2002), p. 11.

10. Hans Jurgen Teuteberg and Jean-Louis Flandrin, "The Transformation of the European Diet," pp. 442–56 in J. Flandrin and M. Montanari, eds., *A Culinary History of Food* (New York: Columbia University Press, 1999); Kathleen DesMaisons, *Potatoes Not Prozac* (New York: Simon & Schuster, 1999); Janice M. Horowitz, "Ten Foods That Pack a Wallop," *Time,* January 21, 2002, p. 81. See also Richard Wurtman and Judith Wurtman, "The Use of Carbohydrate-Rich Snacks to Modify Mood State," pp. 151–56 in G. Harvey Anderson and Sidney Kennedy, eds., *The Biology of Feast and Famine* (New York: Academic Press, 1992); Judith Wurtman, *The Serotonin Solution* (New York: Fawcett Books, 1997).

11. M. F. K. Fisher, *The Art of Eating* (New York: John Wiley & Sons, 1990), p. 23.

12. Stephen Hilgartner, *Science on Stage* (Stanford, CA: Stanford University Press, 2000), p. 29.

13. Ibid.: 9–20. See also Marion Nestle, *Food Politics* (Berkeley: University of California Press, 2002), pp. 70–72.

14. Edwin McDowell, "Behind the Best Sellers," *New York Times,* September 20, 1981, p. 42.

15. Jane Brody, interview, April 26, 2002, and see her "Personal Health" column about this: *New York Times,* June 11, 2002, p. F8.

16. See, for example, Jane Brody's articles and "Personal Health" columns in the *New York Times* of June 18, 1980; June 24, 1981; September 24, 1986; May 10, 1995; May 14, 2001; June 26, 2001; October 9, 2001; June 11, 2002; July 16, 2002; March 9, 2004. Quote is from Brody's remarks at an Oldways conference, San Diego, California, April 23, 2002.

17. Jane Brody, "Personal Health," *New York Times,* January 18, 1984: C1; and April 4, 1984, p. C1. She is more favorable toward eggs elsewhere, for example in *Jane Brody's Nutrition Book* (New York: Bantam, 1981). On the fall and rise of the egg more generally, see Joyce Hendley, "Cracking the Case," *Eating Well* (Spring 2003): 25–32.

18. Over time, Brody contradicts her own advice about fats and cholesterol. For examples from the 1980s, see Russell Smith, *The Cholesterol Conspiracy* (St. Louis: Warren H. Green Publishers, 1991), pp. 28–29.

19. Frank Hu et al., "A Prospective Study of Egg Consumption and Risk of Cardiovascular Disease," *Journal of the American Medical Association* 281 (1999): 1387–94; Stephen B. Kritchevsky and D. Kritchevsky, "Egg Consumption and Coronary Heart Disease," *Journal of the American College of Nutrition* 19 (2000): 549S–555S; Harold McGee, *On Food and Cooking* (New York: Macmillan, 1984), chap. 2.

20. U.S. Food and Drug Administration, "Guidance on How to Understand and Use the Nutrition Facts Panel on Food Labels" (http://vm.cfsan.fda.gov/~dms/foodlab.html); David Kritchevsky, "Diet and Artherosclerosis," *Journal of Nutrition, Health, and Aging* 5 (2001): 155–59; Pirjo Pietinen et al., "Intake of Fatty Acids and Risk of Coronary Heart Disease in a Cohort of Finnish Men," *American Journal of Epidemiology* 145 (1997): 876–87; Uffe Ravnskov, "Diet–Heart Disease Hypothesis Is Wishful Thinking," *British Medical Journal* 324 (2002): 238; Hester Vorster et al., "Egg Intake Does Not Change Plasma Lipoprotein and Coagulation Profiles," *American Journal of Clinical Nutrition* 55 (1992): 400–410; Paul N. Hopkins, "Effects of Dietary Cholesterol on Serum Cholesterol: A Meta-Analysis and Review," *American Journal of Clinical Nutrition* 55 (1992): 1060–70; Esther Lopez-Garcia, M. Schulze, et al., "Consumption of Trans Fatty Acids Is Related to Plasma Biomarkers of Inflammation and Endothelial Dysfunction," *Journal of Nutrition* 135 (2005): 562–66.

21. Emily Green, "No—Less Is Less," *Los Angeles Times,* May 10, 2000.

22. Emily Green, "The Low-Fat-Free, Diet-Food-Free Diet" and "Virtue with a Touch of Gloss," *Los Angeles Times,* March 13 and July 24, 2002.

23. Estimate is based on data from the U.S. Department of Agriculture.

24. See David Kritchevsky, "Antimutagenic and Some Other Effects of

Conjugated Linoleic Acid," *British Journal of Nutrition* 83 (2000): 459–65; Rosaleen Devery et al., "Conjugated Linoleic Acid and Oxidative Behaviour in Cancer Cells," *Biochemical Society Transactions* 29 (2001): 341–44; Marianne O'Shea et al., "Milk Fat Conjugated Linoleic Acid (CLA) Inhibits Growth of Human Mammary MCF-7 Cancer Cells," *Anticancer Research* 20 (2000): 3591–601; Jean-Michel Gaullier, J. J. Halse, et al., "Supplementation with Conjugated Linoleic Acid for 24 Months Is Well Tolerated by and Reduces Body Fat Mass in Healthy, Overweight Humans," *Journal of Nutrition* 135 (2005): 778–84. For additional studies on CLA, see www.wisc.edu/fri/clarefs.htm. Contrary to the impression given by advocates of the gospel of naught, whether low-fat diets are optimal for reducing the risk of heart disease and related outcomes remains an open question. See for example Kulwara Meksawan, David R. Pendergast, et al., "Effect of Low and High Fat Diets on Nutrient Intakes and Selected Cardiovascular Risk Factors in Sedentary Men and Women," *Journal of the American College of Nutrition* 23 (2004): 131–40.

25. Paul Rozin et al., "Lay American Conceptions of Nutrition," *Health Psychology* 15 (1996): 438–47. See also Michael E. Oakes, "Suspicious Minds: Perceived Vitamin Content of Ordinary and Diet Foods with Added Fat, Sugar or Salt," *Appetite* 43 (2004): 105–8.

26. Jane Brody, "High-Fat Diet: Count Calories and Think Twice," *New York Times*, September 10, 2002.

27. Gary Taubes, "What If It's All Been a Big Fat Lie?" *New York Times*, July 7, 2002; Sally Squires, "Into Our Stomachs and Out of Our Minds," "Experts Declare Story Low on Saturated Facts," and "The Skinny on Author Gary Taubes," *Washington Post*, July 28, 2002, and August 27, 2002. For Taubes's response to Squires, see Gary Taubes, "Dietary Fat, Cont'd.," *Washington Post*, September 24, 2002.

28. Publication dates for the Kolata stories: "Benefit . . . ," April 25, 1995; "Amid inconclusive . . . ," May 10, 1995; "In Public Health . . . ," April 23, 2002; "Scientists . . . ," April 30, 2002; "Body Heretic . . . ," April 17, 2005.

29. Quote is from Kolata, "In Public Health. . . ."

30. Kolata, "In Public Health . . ."; John P. A. Ioannidis, "Contradicted and Initially Stronger Effects in Highly Cited Clinical Research," *Journal of the American Medical Association* 294 (2005): 218–28. On serious shortcomings in the Nurses Study and those of similar design, see citations in subsequent footnotes below, and specifically, on the exclusion of subjects from the researchers' analyses, see Debbie A. Lawlor, George Davey Smith, and S. Ebrahim, "The Hormone Replacement–Coronary Heart Disease Conundrum," *International Journal of Epidemiology* 33 (2004): 464–67; Debbie A. Lawlor, George Davey Smith, et al., "Those Confounded Vitamins," *Lancet* 363 (2004): 1724–26; Gina Kolata, "Study Tying Longer Life to Extra Pounds Draws Fire," *New York Times*, May 27, 2005.

31. Moreover, except for smoking and blood pressure, the risk factors for heart disease vary between countries. See, for example, Alessandro Menotti et al., "Cardiovascular Risk Factors as Determinants of 25-Year All-Cause Mortality in the Seven Countries Study," *European Journal of Epidemiology* 17 (2001): 337–46; John Powles, "Commentary: Mediterranean Paradoxes Continue to Provoke," *International Journal of Epidemiology* 30 (2001): 1076–77; and papers presented at the NATO research workshop conference, "Coronary Heart Disease in Central and Eastern Europe," Budapest, May 2000 (www.sote .hu/magtud/ind_nato.htm). See also Gina Kolata, "Eating Well Can't Hurt. But for Now, Scientists Say, the Benefits Remain Hypothetical and Elusive," *New York Times*, September 27, 2005.

32. On the limitations of the research designs and interpretations of findings, see Gary Taubes, "Epidemiology Faces Its Limits," *Science* 269 (1995): 164–69; Gina Kolata, "In Public Health, Definitive Data and Results Can Be Elusive," *New York Times*, April 23, 2002; Beverly Rockhill et al., "Individual Risk Prediction and Population-Wide Disease Prevention," *Epidemiologic Reviews* 22 (2000): 176–80. In a textbook, Walter Willett notes limitations of various designs as well: Walter Willett, *Nutritional Epidemiology* (New York: Oxford University Press, 1998), chap. 1.

33. Willett quote appears in Ellen Ruppel Shell, "Food Warrior," *Technology Review* (November 1996): 33–40. The press coverage and the study are discussed in Jack Hart, "Our Days Are Numbered," *Editor and Publisher* (July 17, 1999): 63. The study: Walter Willett et al., "Relation of Meat, Fat, and Fiber Intake to the Risk of Colon Cancer in a Prospective Study among Women," *New England Journal of Medicine* 323 (1990): 1664–72. (Note that Hart's calculations of relative risks are computed incorrectly.) On limitations of the Nurses Study and those of similar design, see notes just above and below, and my related discussion in chap. 7.

34. Peter Jaret, "The Capsaicin Quandary," *Eating Well* (Summer 2002): 22–26. The original study is Lizbeth Lopez-Carrillo, M. Hernandez Avila, and R. Dubrow, "Chili Pepper Consumption and Gastric Cancer in Mexico," *American Journal of Epidemiology* 139 (1994): 263–71.

35. Georg Simmel, "The Sociology of the Meal," *Food and Foodways* 5 (1994): 345–50 (quote is on p. 346).

36. On problems with diet surveys, see also Nicholas Day et al., "Epidemiological Assessment of Diet," *International Journal of Epidemiology* 30 (2001): 309–17; Shell, "Food Warrior"; Frances A. Larkin, H. L. Metzner, et al., "Comparison of Estimated Nutrient Intakes by Food Frequency and Dietary Records in Adults," *Journal of the American Dietetic Association* 89 (1989): 215–23; Robert C. Klesges, L. H. Eck, and J. W. Ray, "Who Underreports Dietary Intake in a Dietary Recall?" *Journal of Consulting Clinical Psychology* 63 (1995): 438–44; Janet A. Novotny, W. V. Rumpler, et al., "Personality Characteristics as Predictors of Underreporting of Energy Intake on 24-Hour

Dietary Recall Interviews," *Journal of the American Dietetic Association* 103 (2003): 1146–51; Janet A. Tooze and Amy F. Subar, "Psychosocial Predictors of Energy Underreporting in a Large Doubly Labeled Water Study," *American Journal of Clinical Nutrition* 79 (2004): 795–804; Carmen D. Samuel-Hodge, L. M. Fernandez, et al., "A Comparison of Self-Reported Energy Intake with Total Energy Expenditure Estimated by Accelerometer and Basal Metabolic Rate in African-American Women with Type 2 Diabetes," *Diabetes Care* 27 (2004): 663–69.

37. Jaret, "The Capsaicin Quandary"; Mary Ward and Lizbeth Lopez-Carrillo, "Dietary Factors and the Risk of Gastric Cancer in Mexico City," *American Journal of Epidemiology* 149 (1999): 925–32.

38. Felipe Fernandez-Armesto, *Near a Thousand Tables* (New York: Free Press, 2002): 46; see also Harvey Levenstein, *Revolution at the Table* (New York: Oxford University Press, 1988), chap. 16.

39. James Le Fanu, *The Rise and Fall of Modern Medicine* (New York: Carroll & Graf, 1999); Uffe Ravnskov, *The Cholesterol Myths* (Washington, D.C., New Trends Publishing, 2000). Ravnskov also maintains a Web site (www.ravnskov.nu/cholesterol.htm) and a discussion list. John Powles, "Commentary: Mediterranean Paradoxes Continue to Provoke," *International Journal of Epidemiology* 30 (2001): 1076–77; John Powles, review of Gene Spiller, *The Mediterranean Diets in Health and Disease, Health Transition Review* 2 (1992): 115–116; David Kritchevsky, "Diet and Atherosclerosis," *Journal of Nutrition, Health, and Aging* 5 (2001): 155–59.

40. Quote is from Marcia Angell and Jerome Kassirer, "Clinical Research: What Should You Believe?" *Consumers' Research Magazine* (September 1994): 20–21.

41. As we will see, Hahn's view is more widely shared within the medical and research establishment than many food activists and "health beat" journalists would have us believe. As Barnett Kramer, a deputy director of the National Institutes of Health, put it to Gina Kolata of the *New York Times*, "Over time, the messages on diet and cancer have been ratcheted up until they are almost co-equal with the smoking messages. I think a lot of the public is completely unaware that the strength of the message is not matched by the strength of the evidence." Kolata, "Eating Well Can't Hurt. . . ."

42. See, for example, Ronald Krauss, "Dietary and Genetic Effects on Low-Density Lipoprotein Heterogeneity," *Annual Review of Nutrition* 21 (2001): 283–95; Michael Alderman, "Salt, Blood Pressure, and Health," *International Journal of Epidemiology* 31 (2002): 311–16; Yi Ming Fan, J. T. Salonen, et al., "Hepatic Lipase C-480T Polymorphism Modifies the Effect of HDL Cholesterol on the Risk of Acute Myocardial Infarction in Men," *Journal of Medical Genetics* 41 (2004): e28.

43. Meir Stampfer et al., "Primary Prevention of Coronary Heart Disease

in Women through Diet and Lifestyle," *New England Journal of Medicine* 343 (2000): pp. 16–22.

44. Actually, I'm protected even more. The paper goes on to show an additional 31 percent reduction in risk for those of us who drink a moderate amount of alcohol every day.

45. Karin Michels and Alicja Wolk, "A Prospective Study of Variety of Healthy Foods and Mortality in Women," *International Journal of Epidemiology* 31 (2002): 847–54.

46. See, for example, Tone Bjorge et al., "Human Papillomavirus Infection as a Risk Factor for Anal and Perianal Skin Cancer in a Prospective Study," *British Journal of Cancer* 87 (2002): 61–64; Jose Maria Pajares, "Helicobacter Pylori Infection and Gastric Cancer in Spain," *Hepato-Gastroenterology* 48 (2001): 1556–59; Li Li, Emmanuel Messas, et al., "Porphyromonas Gingivalis Infection Accelerates the Progression of Atherosclerosis in a Heterozygous Apolipoprotein E-Deficient Murine Model," *Circulation* 105 (2002): 861–67; J. Thomas Grayston, "Secondary Prevention Antibiotic Treatment Trials for Coronary Artery Disease," *Circulation* 102 (2000): 1742; Russell Ross, "Atherosclerosis—An Inflammatory Disease," *New England Journal of Medicine* 340 (1999): 115–26; Michael Marmot, "Multilevel Approaches to Understanding Social Determinants," pp. 349–67 in L. Berkman and I. Kawachi, eds., *Social Epidemiology* (Oxford: Oxford University Press, 2000); Irvin Emanuel, Weny Leisenring, et al., "The Washington State Intergenerational Study of Birth Outcomes," *Paediatric and Perinatal Epidemiology* 13 (1999): pp. 352–69; Terrence Hill, Catherine Ross, and Ronald Angel, "Neighborhood Disorder, Psychological Distress, and Health," *Journal of Health and Social Behavior* 46 (2005): 170–86; David J. P. Barker, *Mothers, Babies, and Disease in Later Life* (London: British Medical Journal Publishing Group, 1994); Karine Spiegel et al., "Impact of Sleep Debt on Metabolic and Endocrine Function," *Lancet* 354 (1999): 1435–39; Naomi L. Rogers, Martin P. Szuba, et al., "Neuroimmunologic Aspects of Sleep and Sleep Loss," *Seminars in Clinical Neuropsychiatry* 6 (2001): 295–307; Paul Martin, *Counting Sheep* (London: HarperCollins, 2002).

47. In addition to research by Kawachi's group (cited above and below), see Stephen Stansfeld and Michael Marmot, *Stress and the Heart* (London: British Medical Journal Books, 2002); David P. Phillips, George C. Liu, et al., "The Hound of the Baskervilles Effect," *British Medical Journal* 323 (2001): 1443–46; Benjamin A. Shaw, Neal Krause, et al., "Emotional Support from Parents Early in Life, Aging, and Health," *Psychology and Aging* 19 (2004): 4–12; Edmond Shenassa, "Society, Physical Health, and Modern Epidemiology," *Epidemiology* 12 (2001): 467–70; Richard Wilkinson and Michael Marmot, *Social Determinants of Health* (Copenhagen: World Health Organization, 1998); Michael Marmot, *Status Syndrome* (New York: Times

Books, 2004); Michael Marmot, "Life at the Top," *New York Times,* February 27, 2005.

48. Alistair Woodward and Ichiro Kawachi, "Why Should Physicians Be Concerned about Health Inequalities?" *Western Journal of Medicine* 175 (2001): 6–7; Patricia M. Eng, Eric B. Rimm, et al., "Social ties and change in social ries in relation to subsequent total and cause-specific mortality and coronary heart disease incidence in men," *American Journal of Epidemiology* 155 (2002): 700–709; Lisa Berkman, "The Role of Social Relations in Health Promotion," pp. 171–83 in I. Kawachi, B. Kennedy, and R. Wilkinson, eds., *The Society and Population Health Reader* (New York: New Press, 1999); Tores Theorell, "Working Conditions and Health," pp. 95–117 in L. Berkman and I. Kawachi, eds., *Social Epidemiology* (Oxford: Oxford University Press, 2000); Ichiro Kawachi, Norman Daniels, et al., "Health Disparities by Race and Class," *Health Affairs* 24 (2005): 343–53; Jo C. Phelan, Bruce G. Link, et al., "Fundamental Causes of Social Inequalities in Mortality," *Journal of Health and Social Behavior* 45 (2004): 265–86. See also Cheryl K. Nordstrom, Kathleen M. Dwyer, et al., "Work-Related Stress and Early Atherosclerosis," *Epidemiology* 12 (2001): 180–85; Beverly Rockhill, "The Privatization of Risk," *American Journal of Public Health* 91 (2001): 365–68; Mesfin Mulatu and Carmi Schooler, "Causal Connections between Socio-Economic Status and Health," *Journal of Health and Social Behavior* 43 (2002): 22–41; Eliza K. Pavalko, Krysia N. Mossakowski, and Vanessa J. Hamilton, "Does Perceived Discrimination Affect Health?" *Journal of Health and Social Behavior* 44 (2003): 18–33.

49. Ichiro Kawachi and Lisa Berkman, "Social Cohesion, Social Capital, and Health," pp. 174–90 in Lisa Berkman and Ichiro Kawachi, eds., *Social Epidemiology* (Oxford: Oxford University Press, 2000); Bruce Kennedy et al., "Income Distribution and Mortality," "Income Distribution, Socioeconomic Status, and Self-Rated Health," "(Dis)respect and Black Mortality," and "Women's Status and the Health of Women and Men," pp. 60–68, 137–47, 465–73, and 474–91 in I. Kawachi, B. Kennedy, and R. Wilkinson, eds., *The Society and Population Health Reader* (New York: New Press, 1999). See also Edmond Shenassa, "Society, Physical Health, and Modern Epidemiology," *Epidemiology* 12 (2001): 467–70. For additional evidence, see Paula M. Lantz, John W. Lynch, et al., "Socioeconomic Disparities in Health Change in a Longitudinal Study of U.S. Adults," *Social Science and Medicine* 53 (2001): 29–40; Ana Diez Roux, "Neighborhood of Residence and Incidence of Coronary Heart Disease," *New England Journal of Medicine* 345 (2001): 99–106; Redford Williams, John Barefoot, and Neil Schneiderman, "Psychosocial Risk Factors for Cardiovascular Disease," *Journal of the American Medical Association* 290 (2003): 2190–92.

50. J. Fraser Mustard and John Frank, "Determinants of Health from a Historical Perspective," *Daedalus* 123 (1994): 1–19; Richard Wilkinson,

"Health Inequalities," Lisa Berkman, "The Role of Social Relations in Health Promotion," and Bruce McEwen, "Protective and Damaging Effects of Stress Mediators," pp. 148–54, 171–83, and 379–92 in I. Kawachi, B. Kennedy, and R. Wilkinson, eds., *The Society and Population Health Reader* (New York: New Press, 1999).

Chapter 2: Safe Treyf

1. Gaye Tuchman and Harry Levine, "New York Jews and Chinese Food: The Social Construction of an Ethnic Pattern," pp. 163–84 in Barbara Shortridge and James Shortridge, eds., *The Taste of American Place* (New York: Rowman & Littlefield, 1997).

2. Mike Duff, " 'Other White Meat' Marketing Has Pork Living High on the Hog," *DSN Retailing Today,* February 5, 2001.

3. Jonathan Gold, "Spamming the Globe," pp. 271–73 in Holly Hughes, ed., *Best Food Writing 2000* (New York: Marlowe, 2000); Lori Dahm, "Foolproof and Flawless: Hormel's Always Tender Products Offer Consumers a Perfected Meal Solution," *New Products Magazine* (August 2002): 18–19.

4. Bonnie Liebman and Jayne Hurley, "The Best and Worst Fast Food 2002," *Nutrition Action Healthletter* (September 2002): 12–14. Burger King extended its chicken offerings in subsequent years to include a TenderCrisp line of sandwiches. On this, and the chicken trend in fast-food chains more generally, see Theresa Howard, "Fast-Food Firms Hope to Get More Bucks with Clucks," *USA Today,* July 28, 2004.

5. National Consumers League survey, January 2002 (available at www .nclnet.org/naturalsreport.pdf); Daniel Harris, "The Natural," *Salmagundi* (Spring 2000): 236–51; David P. Barash, "The Tyranny of the Natural," *Chronicle of Higher Education,* November 2, 2001, pp. B16–17.

6. Eric Schlosser, *Fast Food Nation* (Boston: Houghton Mifflin, 2001), p. 127.

7. FDA Office of Regulatory Affairs, Compliance Policy Guides Manual, subchapter 555, p. 266.

8. Lori Dahm, "Flavor Spotlight," *New Products Magazine,* January 2003; Food and Drug Administration Code of Regulations (revised April 2002), title 21, section 101, part 22 (available at http://vm.cfsan.fda.gov/~lrd/CF101-22. html).

9. "Naturally Misleading: Consumers' Understanding of 'Natural' and 'Plant-Derived' Labeling Claims," a report from the National Consumers League, 2002 (posted at its Web site); Shane Starling, "Manufacturers Favour Natural Flavours," *Functional Foods & Nutraceuticals,* (March 2002): 30–32.

10. See Marion Nestle, *Food Politics* (Berkeley: University of California Press, 2002), chap. 13; Brian Steinberg, "Food Makers Playing Up Nutrition," *Wall*

Street Journal, March 26, 2004; Deborah Ball, "With Food Sales Flat, Nestle Stakes Future on Healthier Fare," *Wall Street Journal,* March 18, 2004; Margaret Webb Pressler, "The New Apple a Day," *Washington Post,* June 26, 2004; Melanie Warner, "Eating Your Way to Health," *New York Times,* December 28, 2005. The figure $20 billion is based on statistics from trade groups.

11. Elspeth Probyn, *Carnal Appetites* (New York: Routledge, 2000), p. 15.

12. Scott Bruce and Bill Crawford, *Cerealizing America* (Boston: Faber & Faber, 1995), chap. 3.

13. Nestle, *Food Politics,* p. 296. The figure 100 million is derived from surveys by the International Food Information Council.

14. Larry Becker, "Food Fads and You," *Vibrant Life* (June 1999): 3.

15. "Carcinogens Found in Nation's Water," *Washington Post,* March 20, 1977; "Sales Boon for Bottled Water," *New York Times,* August 8, 1982; M. Moore, "Sales Are Bubbling," *USA Today,* May 30, 1989.

16. Rob Walker, "A Spoonful of Attitude," *New York Times Magazine,* August 22, 2004; Sherri Day, "Bottled Water Is Still Pure, but It's Not Simple Anymore," *New York Times,* August 3, 2002; Barbara Durr, "Americans Get Taste for Bottled Water," *Financial Times,* June 24, 2002; Carol Ness, "Vitamin W," *San Francisco Chronicle,* April 2, 2003; Robert Davis, "Aches & Claims," *Wall Street Journal,* November 12, 2002; Theresa Howard, "Enhanced Waters Pour onto Shelves," *USA Today,* August 23, 2002; Greg Kitzmiller, "Examining the Market for Nutraceutical Water," *Nutraceuticals World* (August 2002); and information from the manufacturers' Web sites.

17. Mark Messina, Johanna W. Lampe, et al., "Reductionism and the Narrowing Nutrition Perspective," *Journal of the American Dietetic Association* 101 (2001): pp. 1416–19; Michael Pollan, "The Futures of Food," *New York Times,* May 4, 2003; Nestle, *Food Politics,* part 5; Melinda Fulmer, "What Comes First, the Egg or the Algae DHA Omega-3?" *Los Angeles Times,* September 8, 1999, pp. A1 and A9; Eric Peterson, "OmegaTech Getting Its Eggs in a Row," *Boulder Country Business Report,* December 17, 1999; Katy McLaughlin, "Shelling Out for Designer Eggs," *Wall Street Journal,* March 17, 2004.

18. A. L. Neuhofel, J. H. Wilton, et al., "Lack of Bioequivalence of Ciprofloxacin When Administered with Calcium-Fortified Orange Juice," *Journal of Clinical Pharmacology* 42 (2002): 461–66. The study does not rule out, however, the possibility that the orange juice itself had an effect.

19. "Fortified Foods May Make Certain Drugs Less Effective," *Tufts Health & Nutrition Letter* (December 2002): 8; Debbie Howell, "Vitamins Add Powerful Punch to Keep Juice Sales Flowing," *Discount Store News,* March 20, 2000; "Fortified Foods: Too Much of a Good Thing?" *Consumer Reports,* October 2003; Jane Spencer, "The Risks of Mixing Drugs and Herbs," *Wall Street Journal,* June 22, 2004.

20. The Food Marketing Institute and Rodale survey (cited at www.

newhope.com/nfm-online/nfm_backs/Jan_02/commingle_s1.cfm). See also Warner.

21. Matthew Grimm, "Veggie Delight," *American Demographics* (August 2000): 66–67. Session titles are from "Functional Foods 2002" conference held in San Diego in July 2002. These and similar terms are used also in trade magazines and by organizations such as the Food Marketing Institute and the Natural Marketing Institute (see their "Health and Wellness Trends" reports). See also Linda Gilbert, "Marketing Functional Foods: How to Reach Your Target Audience," *AgBioForum* 3 (2000): 20–38.

22. I checked the Web sites soon after Altria purchased the lines, and again two years later. Information about the content of sessions comes also from observations at functional-foods industry conferences I attended.

23. On gender-specific food products, see also William Roberts, "Gender Vending," *Prepared Foods,* July 2004.

24. Piero Camporesi, *The Magic Harvest* (Cambridge, UK: Polity Press, 1998), p. 15; Judith B. Jones, "A Religious Art," pp. 38–41 in Daniel Halpern, ed., *Not for Bread Alone* (New York: Ecco, 1993); Felipe Fernandez-Armesto, *Near a Thousand Tables* (New York: Free Press, 2002), chap. 4.

25. Bruce and Crawford, *Cerealizing America,* pp. 235–40; Nestle, *Food Politics,* pp. 322–24; Joanna F. Swain, Ian L. Rouse, et al., "Comparison of the Effects of Oat Bran and Low-Fiber Wheat on Serum Lipoprotein Levels and Blood Pressure," *New England Journal of Medicine* 322 (1990): 147–52; Duncan Campbell, "Junk Food Firms Fear Being Eaten Alive by Fat Litigants," *Guardian* (London), May 24, 2002; David Bosworth and Rochelle Gurstein, "The Science of Self-Deception," *Salmagundi* (Fall 1999). The "scientifically proven" quote is from Quaker's Web site. Newspaper headlines: Kevin Cowherd, "The Elixir of Living a Longer Life," *Baltimore Sun,* December 20, 1988; "Oat Bran: Preventative Medicine," *Los Angeles Times,* April 24, 1988; Tony Korenheiser, "Oat Bran Bites the Dust," *Washington Post,* January 19, 1990; Carole Sugarman, "The Rise and Fallacies of Oat Bran," *Washington Post,* January 31, 1990.

26. Judie Diezak, "Demystifying Label Claims," *Prepared Foods* (June 2002); Gary Ford, "The Effects of the New Food Labels on Consumer Decision Making," *Advances in Consumer Research* 21 (1994): 530; Nestle, *Food Politics,* chaps. 10 and 14. Subsequent to the publication of Nestle's book, the food industry got the FDA to loosen its regulations still further: Claudia O'Donnell, "Delivering Heart Health," *Prepared Foods* (July 2004); Vicki Kemper, "FDA Will Let Food Industry Put Health Claims on Labels," *Los Angeles Times,* July 11, 2003; Norm Alster, "New Food Labels Winning Few Fans," *New York Times,* March 28, 2004. See also Martijn Katan and Nicole de Roos, "Toward Evidence-Based Health Claims for Foods," *Science* 299 (2003): 206–7.

27. Stephanie Thompson, "Food for Thought: Companies Seek to Create Buzz with PR, Ads Capitalizing on Nutrition Boom," *Advertising Age,* February 28, 2000, pp. 20–22; David Hendee, "Where's the Beef?" *Meat & Poultry* (Octo-

ber 2002): 26–30; "Superior Healthy Cup Mocha Latte Offers Health Benefits in a Hot Beverage," *Restaurant Show Daily,* May 19, 2002, p. 59; Laurie Tarkan, "As a Hormone Substitute, Soy Is Ever More Popular, but Is It Safe?" *New York Times,* August 24, 2004.

28. Marian Burros, "Doubts Cloud Rosy News on Soy," *New York Times,* January 26, 2000; ABC News, *20/20,* June 9, 2000; Nestle, *Food Politics,* p. 328; Joyce Hendley, "The Enigmatic Bean," *Eating Well* (Summer 2003); Tarkan, "As a Hormone Substitute."

29. Oksana A. Matvienko, Douglas S. Lewis, et al., "A Single Daily Dose of Soybean Phytosterols in Ground Beef Decreases Serum Total Cholesterol and LDL Cholesterol in Young, Mildly Hypercholesterolemic Men," *American Journal of Clinical Nutrition* 76 (2002): 57–64; John O'Neil, "Soy Burgers That Keep the Beef," *New York Times,* June 25, 2002. And regarding ConAgra, see Schlosser, *Fast Food Nation,* pp. 157–60 and 219.

30. Janet Walzer, "This 'Wonder Food' May Be More Hype Than Help," *Tufts Journal* (September 2002); *Tufts Nutrition Notes* (December 2002).

31. Sanne Kreijkamp-Kaspers, Linda Kok, et al., "Effect of Soy Protein Containing Isoflavones on Cognitive Function, Bone Mineral Density, and Plasma Lipids in Postmenopausal Women," *Journal of the American Medical Association* 292 (2004): 65–74.

32. Burros, "Doubts"; Colin Berry, "Risk, Science, and Society," posted at www.spiked-online.com/Articles/00000002D29C.htm. See also Judy Foreman, "Debate Is Intensifying As the FDA Reviews Claims That the Protein May Reduce the Risk of Certain Types of Cancer," *Los Angeles Times,* June 14, 2004.

33. See Tarkan, "As a Hormone Substitute"; Foreman, "Debate Is Intensifying"; and "Soy Bean Hastening Amazon Destruction," Reuters, June 27, 2003. See also "Ranchers, Soybean Farmers, and Loggers Destroyed 10,088 Square Miles of Amazon Rain Forests in 2004," Associated Press, May 19, 2005.

34. Titles are from programs for Soyfoods Summits (2003–2005) and for the International Symposium (available at iqpc.com and aocs.org). Other information is from mailings to journalists from the Soyfoods Council about its Soyfoods Food Editor Tours.

35. On blueberry marketing, see also Barbara Carton, "With Wild Blueberries on the Verge of Glut, a Hunt for New Uses," *Wall Street Journal,* April 23, 2003.

36. "Food Additives—What You Need to Know," *UC Berkeley Wellness Letter* (June 2002).

37. Laura Mason, "Chocolate," pp. 176–81 in Alan Davidson, ed., *The Oxford Companion to Food* (New York: Oxford University Press, 1999).

38. Norman Hollenberg, "Potential Heart Health Benefits of Cocoa Polyphenols," posted on www.terepia.com.

39. Sabra Chartrand, "Patents," *New York Times,* February 11, 2002. Health and nutrition benefits: Ying Wan et al., "Effects of Cocoa Powder and Dark

Chocolate on LDL Oxidative Susceptibility and Prostaglandin Levels in Humans," *American Journal of Clinical Nutrition* 74 (2001): 596–602; Thomas H. Parliament, Chi-Tang Ho, and Peter Schieberle, eds., *Caffeinated Beverages: Health Benefits, Physiological Effects, and Chemistry* (New York: Oxford University Press/American Chemical Society, 2000); Robert R. Holt, Derek D. Schramm, et al., "Flavonoid-Rich Chocolate and Platelet Function," *Journal of the American Medical Association* 287 (2002): 2212–13; Marion Nestle, "The Role of Chocolate in the American Diet," pp. 111–24 in Alex Szogyi, ed., *Chocolate: Food of the Gods* (Westport, CT: Greenwood, 1997); Mauro Serafini, Rossana Bugianesi, et al., "Plasma Antioxidants from Chocolate," *Nature* 424 (2003): 1013; Penny Kris-Etherton, Carl Keen, et al., "Evidence That the Antioxidant Flavonoids in Tea and Cocoa Are Beneficial for Cardiovascular Health," *Current Opinion in Lipidology* 13 (2002): 41–49; Mary B. Engler, Marguerite M. Engler, et al., "Flavonoid-Rich Dark Chocolate Improves Endothelial Function and Increases Plasma Epicatechin Concentrations in Healthy Adults," *Journal of the American College of Nutrition* 23 (2004): 197–204; Qin Yan Zhu, Derek D. Schramm, et al., "Influence of Cocoa Flavonols and Procyanidins on Free Radical–Induced Human Erythrocyte Hemolysis," *Clinical & Developmental Immunology* 12 (2005): 27–34; Davide Grassi, Cristina Lippi, et al., "Short-Term Administration of Dark Chocolate Is Followed by a Significant Increase in Insulin Sensitivity and a Decrease in Blood Pressure in Healthy Persons," *American Journal of Clinical Nutrition* 81 (2005): 611–14. See also Elizabeth Olson, "Beyond Delicious, Chocolate May Help Pump Up Your Heart," *New York Times,* February 17, 2004; Jon Gertner, "Eat Chocolate, Live Longer?" *New York Times,* October 10, 2004.

40. On mood effects, see for example Katri Raikkonen, Anu K. Pesonen, et al., "Sweet Babies: Chocolate Consumption During Pregnancy and Infant Temperament at Six Months," *Early Human Development* 76 (2004): 139–45.

41. Jeffrey Steingarten, *The Man Who Ate Everything* (New York: Knopf, 1997), p. 209, and *It Must've Been Something I Ate* (New York: Knopf, 2002), pp. 177–86.

Chapter 3: Promises of the Fathers

1. See Joan Dye Gussow, "The Incompatibility of Food and Capitalism," *Snail* (August 2002): 22–25.

2. The advertising copy is from Walnut Acres.

3. Diane Bourn and John Prescott, "A Comparison of the Nutritional Value, Sensory Qualities, and Food Safety of Organically and Conventionally Produced Foods," *Critical Reviews in Food Science & Nutrition* 42 (2002): 1–34; Melissa Healy, "Behind the Organic Label," *Los Angeles Times,* September 6, 2004; Eleena de Lisser, "Is That $5 Gallon of Milk Really Organic?" *Wall Street Journal,* August 20, 2002; "Pesticides and Kids' Risks," *Newsweek,* June

1, 1998; Scott McCredie, "Organic Produce Is Growing More Popular, but Is It Really the Healthier Choice?" *Seattle Times,* July 31, 2002; Anthony Trewavas, "A critical assessment of organic farming and food assertions," *Crop Protection* 23 (2004): 757–781; K. Woese et al., "A Comparison of Organically and Conventionally Grown Foods," *Journal of the Science of Food and Agriculture* 74 (1997): 281–93; Terence Hines, "Do Pesticides Cause Cancer?" *Skeptic Magazine* 10, no. 4 (2004): 26–27; Maribel I. Fernandez and Brent W. Woodward, "Comparison of Conventional and Organic Beef Production Systems," *Livestock Production Science* 61 (1999): 213–23; Allan S. Felsot and Joseph D. Rosen, "Comment on Comparison of the Total Phenolic and Ascorbic Acid Content of Freeze-Dried and Air-Dried Marionberry, Strawberry, and Corn Grown Using Conventional, Organic, and Sustainable Agricultural Practices," *Journal of Agriculture and Food Chemistry* 52 (2004): 146–49. ADA and USDA information is from their Web sites. Note also that organic farming entails environmental costs as well as benefits. See, for example, Blake Hurst, "Up on the Farm," *Wilson Quarterly* (Summer 2003): 42–51.

4. J. I. Rodale, *Organic Gardening* (Emmaus, PA: Rodale Books, 1955); Harvey Levenstein, *Paradox of Plenty* (New York: Oxford University Press, 1993), pp. 162 and 181; Shelley Davis, "Doing What Comes Naturally," *Washington Post,* July 15, 1984; Wade Greene, "Guru of the Organic Food Cult," *New York Times Magazine,* June 6, 1971; and information on Rodale posted at rodale.com.

5. See Dan Campbell, "Cream of the CROPP: Demand Outstripping Supply as CROPP Organic Dairy Products Go Prime Time," *Rural Cooperatives* (May–June 2005): 15–18; Shermain Hardesty, "Positioning California's Agricultural Cooperatives for the Future," *Agricultural and Resource Economics Update* 8 (2004): 7–10.

6. See also Vince Beiser, "Harvest of Pain," *LA Weekly,* December 4, 2003.

7. Jane Brody, "A World of Food Choices, and a World of Infectious Organisms," *New York Times,* January 30, 2001.

8. James DeWan, "Zapping Public Fear About Food Irradiation," *Philadelphia Inquirer,* July 10, 2002.

9. Kevin Orland, "Restaurants Serve Irradiated Meat," Associated Press, September 10, 2002; Lori Lohmeyer, "Embers Promotes Irradiated Burgers," *Wisconsin State Journal,* November 11, 2002; Dan Mayfield, "Zap! Chain Brings Electrified Beef to N.M.," *Albuquerque Tribune,* September 17, 2003.

10. Charles Hirshberg, "Examining the Quorntroversy," *Vegetarian Times* (February 2003). See also Joseph P. Lewandowski, "Quorn Dogged," *Natural Foods Merchandiser* (October 2002); Rosie Mestel, "A Food Fight over Fungus," *Los Angeles Times,* March 12, 2004.

11. Denise Grady, "What's in Those Nuggets?" and Eric Asimov, "Have It Your Way," *New York Times,* May 14, 2002.

12. Marian Burros, "A Definition at Last, but What Does It All Mean?" *New York Times*, October 16, 2002; Samuel Fromartz, "Small Organic Farmers Pull Up Stakes," *New York Times*, October 14, 2002; Trish Hall, "Sunnyside Up," *Gourmet* (July 2001): 66–69; Steve Werblow, "Small Organic Farmers Confront New Terrain," *Eating Well* (Spring 2003): 77–78; Max Withers, "The Many Meanings of 'Organic,' " *Los Angeles Times*, May 4, 2005. On faulty reasoning in the "buy local" philosophy, see Patrick Martins, "Set That Apricot Free," *New York Times*, April 24, 2004.

13. Michael Pollan, "Naturally," *New York Times Magazine*, May 13, 2001.

14. See also Wade Greene, "Guru of the Organic Food Cult."

15. On the rapid growth of big food companies' control of organic acreage and production, and continuing debates about "big organic," see Lori Dahm, "Organic Panic," *New Products Magazine* (March 2004): 30–34; Elizabeth Weise, " 'Organic' Milk Needs a Pasture," *USA Today*, March 9, 2005.

16. See also William Roberts, "A Natural Progression," *Prepared Foods* (June 2001): 12–16.

17. "Building in Sensory Perception from the Ground Up," *New Products Magazine*, (January 2003), and information from McCormick.

18. Carolyn Jung, "Who Thinks Up This Stuff?" *San Jose Mercury News*, February 19, 2003.

19. United States Patent 6,007,863, awarded December 28, 1999.

20. For more on Mattson, see Malcolm Gladwell, "The Bakeoff," *New Yorker*, September 5, 2005, pp. 125–33.

21. Quote is from "FDA Backgrounder, May 1999: The Food Label," available at www.cfsan.fda.gov/dms/fdnewlab.html. Other information is from "Paradise Tomato Kitchens, Inc.," NAD/CARU Case Reports December 2001, pp. 473–76; a letter dated August 21, 2000, from Robert Ilse, president of Stanislaus Food Products, to Dockets Management Branch of the FDA, available from the FDA Web site; and information available on the companies' Web sites.

22. Fernando Vallejo, Cristina Garcia-Viguera, et al., "Changes in Broccoli Health-Promoting Compounds with Inflorescence Development," *Journal of Agricultural and Food Chemistry* 51 (2003): 3776–82; "Frozen Veg 'Healthier Than Fresh,' " BBC News, March 31, 2003 (reporting a study by Austrian Consumers' Association, available at www.konsument.at); Minna Morse, "The Freshness Fallacy," *Health* (March 1998): 38–41; Hazel Curry, "When Fresh Isn't Best," *Evening Standard* (London), October 22, 2002; "Canned and Frozen Fruit Provide Winter Nutrition," *USA Today*, October 12, 2002. For the canned-foods industry's rap on this, see www.mealtime.org. And see also Roisin Pill, "An Apple a Day: Some reflecting on working-class mothers' views of food and health," pp. 117–27 in Anne Murcott, ed., *The Sociology of Food and Eating* (London: Gower, 1983).

23. See Margaret Visser, *Much Depends on Dinner* (New York: Grove Press, 1987), chap. 6.

24. Julia Moskin, "Sushi Fresh from the Deep . . . the Deep Freeze," *New York Times,* April 8, 2004. Another example is French fries. Under USDA law, frozen French fries can be labeled as "fresh vegetables" (Andrew Martin, "Frozen Fries Are 'Fresh' Veggies," *Los Angeles Times,* June 15, 2004). On the frequent but unacknowledged use of frozen vegetables by prominent chefs, see Mark Bittman, "Frosty the Vegetable," *New York Times,* February 16, 2005.

25. The postings are from exchanges in March 2003 on the online discussion forum for members of the Association for the Study of Food and Society.

26. Harvey Levenstein, *Revolution at the Table* (New York: Oxford University Press, 1988), chaps. 3 and 8; Leslie Brenner, *American Appetite* (New York: Harper-Collins, 1999), p. 16. See also Giorgio Pedrocco, "The Food Industry and New Preservation Techniques," pp. 481–91 in Jean-Louis Flandrin and M. Montanari, eds., *Food: A Culinary History* (New York: Columbia University Press, 1999).

27. Levenstein, *Paradox of Plenty,* p. 117.

28. Levenstein, *Paradox of Plenty,* p. 111; Sallie Tisdale, *The Best Thing I Ever Tasted* (New York: Riverhead, 2000, p. 183).

29. Lori Dahm, "The Dinner Dilemma," *New Products Magazine* (February 2003): 24–25.

30. "Pretty Krafty," *Nutrition Action Healthletter* (April 2001).

Chapter 4: Restaurant Heaven

1. Ruth Reichl, *Comfort Me with Apples* (New York: Random House, 2001), p. 6.

2. Adam Gopnik, "The Cooking Game," *New Yorker,* August 19, 2002, p. 88.

3. Maile Carpenter, "Eating in Michael Bauer's Town," *San Francisco Magazine* (August 2001); Leslie Brenner, *The Fourth Star* (New York: Clarkson Potter, 2002); Andrew Dornenburg and Karen Page, *Dining Out* (New York: John Wiley, 1998, p. 123 for quotation from Kinkead); Cynthia Cotts, "Amanda Hesser's Spice Market Review Sparks Debate over Foodie Ethics and Tableside Manners," *Village Voice,* April 6, 2004.

4. See also Mitchell Davis, "Power Meal," *Gastronomica* 4 (Summer 2004): 60–72.

5. In some accounts of this famous set of incidents, Reichl was anonymous neither time. In support of his contention that reviewers are oblivious of their lack of anonymity, Mitchell Davis told me, "The joke there is, and I've heard this straight from the maître d's mouth, they knew who she was both times." If that report is accurate, one can only imagine how disappointing the experience must have been for true plebeians there.

6. Frank Bruni, "The Magic of Napa with Urban Polish," *New York Times,* September 8, 2004. Similarly, in awarding four stars to Le Bernardin, he ac-

knowledged both that he was "repeatedly recognized" by the chef and that "I regularly field complaints from friends who found their experiences there disappointing." But he attributed the latter to excessively high expectations by the complainants (Bruni, "Only the Four Stars Remain Constant," *New York Times*, March 16, 2005). Marian Burros, the *Times* food section writer and sometime reviewer, also believes the myth. "Having been spotted at restaurants throughout my reviewing career, I have learned one thing: the owners cannot improve the food for the reviewer's sake," she wrote in one of her reviews ("Tapas for Really Close Friends," *New York Times*, January 28, 2004).

7. Brenner, *The Fourth Star*, p. 314. And note comments like Charlie Trotter's: "I don't want to be the best at what I do. I want to be the only person who does what I do" (Edmund Lawler, *Great Restaurants of the World: Charlie Trotter's* [New York: Lebhar-Friedman Books, 2002]).

8. See for example "World's 50 Best Restaurants," *Restaurant Magazine* (June 2003).

9. Brenner, *The Fourth Star*, pp. 17 and 63. See also Steven Shaw, *Turning the Tables* (New York: HarperCollins, 2005).

10. Brenner, *The Fourth Star*, pp. 41–42; David Shaw, "They Have a File on You," *Los Angeles Times*, April 23, 2003. Many restaurants value OpenTable for another reason. Labor and other costs make phone reservations about four times as costly to the restaurant as OpenTable reservations. "Online and In-Person: Tips for Living Like a VIP," *Wall Street Journal*, June 14, 2005.

11. Mark Bittman, "A Taste of Los Angeles," *New York Times*, May 7, 2003; Amanda Hesser, "The Chef," May 7, 21, and June 4, 2003. On treatment of VIPs and regular patrons, see also Steven Shaw, *Turning the Tables*.

12. Rebecca L. Spang, *The Birth of the Restaurant* (Cambridge, MA: Harvard University Press, 2001) (quote is from p. 223).

13. Spang, chap. 6 (quote is from p. 150).

14. Spang, epilogue (quote is from p. 234).

15. See also Patric Kuh, "Rolling a-a-a-nd Ashton," *Los Angeles Magazine* (September 2003): 128–30.

16. Patric Kuh, *The Last Days of Haute Cuisine* (New York: Viking, 2001), p. 194.

17. William Grimes, "A Big Room for Bigger Appetites," *New York Times*, April 23, 2003.

18. See also Adam Nagourney, "24 Restaurants and Still Hungry," *New York Times*, June 22, 2005.

19. Florence Fabricant and Marian Burros, "Rocco DiSpirito Is Out at Union Pacific," *New York Times*, September 29, 2004.

20. Spang, p. 177.

21. A model-cum-bartender from the cast told a reporter, "Our prize was our exposure. Anyone who works in the entertainment industry—whether they are modeling or acting—wants that exposure." Abby Morris, "Local Woman

Completes Run on NBC Reality Show," *Elizabethton* (TN) *Star,* September 3, 2003.

22. Devin Gordon, "The Knives Are Out," *Newsweek,* July 7, 2003; William Grimes, "Cash. Fame. Pressure. And Garlic," *New York Times,* July 20, 2003; Lewis Beale, "Recipe for Trouble," *Los Angeles Times,* July 20, 2003.

23. Gary Levin, "This 'Restaurant' Hams It Up with Reality Fare," *USA Today,* July 8, 2003.

24. Christian L. Wright, "Get Real," *Gourmet* (September 2003).

25. William Grimes, "Rocco's on 22nd," *New York Times,* July 25, 2003.

26. Kuh, *Last Days.*

27. Craig Claiborne, "Receives Rare Public Enthusiasm," *New York Times,* November 26, 1960 (La Caravelle review).

28. Dorenburg and Page, *Dining Out,* p. 15; Kuh, *Last Days,* pp. 121–48.

29. See also David Pasternack, "The Chef," *New York Times,* July 24, 2002.

30. Leslie Brenner, *American Appetite* (New York: HarperCollins, 2000), pp. 214–15. Or as Adam Gopnik writes, "The old French cooking was all harmony, with the chicken breast or the sole treated as the bass note, and everything else coming from what went on top. The new cooking, which has spread from America and Australia out into the world, is almost purely melodic: the unadorned perfect thing, singing the song of itself" (Adam Gopnik, "Two Cooks," *New Yorker,* September 5, 2005, pp. 91–98, quote appears on p. 98).

31. Martin Booe, "In the Chefs' Secret Service," *Los Angeles Times,* June 4, 2000.

32. Michael Ruhlman, "The Pittsburgh Lamber," p. 195 in Thomas Keller, *The French Laundry Cookbook* (New York: Workman, 1999).

33. Leslie Brenner, *The Fourth Star;* Tom Sietsema, "Secret Ingredients," *Washington Post,* January 12, 2000.

34. Keller, *The French Laundry Cookbook,* pp. 106–7.

35. For some of the key roots of this approach during the 1980s, see Caroline Bates, "When Tokyo Met Lyon, or the Rise and Rise of Asian Fusion," *Gourmet* (June 2005).

36. Daniel Boulud, *Letters to a Young Chef* (New York: Basic Books, 2003), pp. 24–25, 50, 75, 98, 103.

Chapter 5: The Food Adventurers

1. These examples are from postings on www.chowhound.com (Web site discussed below).

2. David Kirby, "Gastronome, Know Thyself," *New York Times,* January 6, 2002.

3. The Fisher quote appears as an epigraph in Richard Sterling, *The Fearless Diner* (San Francisco: Travelers' Tales Press, 1998), and Lisa Heldke, *Exotic*

Appetites (New York: Routledge, 2003). The latter author uses the term "food adventurer" in much the way I do here.

4. Janelle Brown, "Valet Parking? That's So 2000," *New York Times,* August 20, 2003.

5. Jane Dornbusch, "The Chowhound Manifesto," *Boston Herald,* June 27, 2001 (Leff quote).

6. Jesse Katz, "Diner's Club," *Los Angeles Magazine* (September 2004).

7. July 14, 2003, posting on www.chowhound.com's Los Angeles board.

8. "Crepe Show," *City Pages,* November 5, 2003; Gold quotes are from his "Where to Eat" columns in 2003.

9. Jonathan Gold, *Counter Intelligence* (New York: St. Martin's, 2000), p. 232.

10. William Grimes, "Critic's Notebook," and Florence Fabricant, "Little Restaurant Elbows the 4-Stars in Zagat's New List," *New York Times,* October 20, 2003; the Chowhound post was on October 22, 2003; Leff quote is in Dornbusch, "The Chowhound Manifesto."

11. See also Calvin Trillin, *Feeding a Yen* (New York: Random House, 2003), pp. 69–88.

12. Five months after our dinner there, Asimov published a favorable review of the place. Perhaps the quality of the food had improved, or we had been there on an off night. Or maybe he ordered better. Reading Asimov's review, I was struck that the dish he extolled, the paella, was not among those the chowhounds ordered, and that he expressed frustration with other dishes and with what he called the "quirkiness" of the place. Eric Asimov, "Tucked Away, an Unassuming Cantina in Chelsea," *New York Times,* March 17, 2004.

13. "Bad" quote is from a posting by "Kip" on October 14, 2003; Leff post is from March 8, 2002. Surveys by the National Restaurant Association find that people who patronize ethnic restaurants avidly seek out authenticity and consider it at least as important as taste. National Restaurant Association, Ethnic Cuisines I (1994) and Ethnic Cuisines II (2000).

14. See Amy Bentley, review of Richard Pillsbury, *No Foreign Food,* in *Gastronomica* (Winter 2001): 104–6.

15. Heldke, *Exotic Appetites,* pp. 31, 115–17; Jeffrey Pilcher, "From 'Montezuma's Revenge' to 'Mexican Truffles,'" pp. 76–96 in Lucy Long, ed., *Culinary Tourism* (Lexington: University Press of Kentucky, 2004), p. 87; and see, for example, the menu for the "Taste of Two Mexicos" event at the James Beard Foundation, February 11, 2000 (www.jamesbeard.org).

16. Jennie Germann Molz, "Tasting an Imagined Thailand," pp. 53–75 in Lucy Long, ed., *Culinary Tourism* (University Press of Kentucky, 2004). The term "staged authenticity" is derived from Erving Goffman, *The Presentation of Self in Everyday Life* (New York: Doubleday, 1959). See also Shun Lu and Gary Alan Fine, "The Presentation of Ethnic Authenticity: Chinese Food as a Social Accomplishment," *Sociological Quarterly* 36 (1995): 535–53; and for the

related concept "planned authenticity," see Meredith Abarca, "Authentic or Not, It's Original," *Food and Foodways* 12 (2004): 1–25.

17. Laurier Turgeon and Madeleine Pastinelli, " 'Eat the World': Postcolonial Encounters in Quebec City's Ethnic Restaurants," *Journal of American Folklore* 115 (2002): 247–69.

18. Linda Burum, *Ethnic Food in Los Angeles* (New York: HarperCollins, 1992), pp. 77–79; Donna R. Gabaccia, *We Are What We Eat* (Cambridge, MA: Harvard University Press, 1998, p. 2 for Paz quote); Sarah Kershaw, "Borscht, Please, with a Side of Sushi," *New York Times,* December 31, 2001. On how the demand for authenticity can limit culinary evolution and experimentation, see Abarca.

19. Gabaccia.

20. Quote is from Samantha Barbas, "I'll Take Chop Suey: Restaurants as Agents of Culinary and Cultural Change," *Journal of Popular Culture* 36 (2003): 669–80. The range of ethnic places was considerably more limited prior to the Immigration and Nationality Act of 1965, a piece of legislation Calvin Trillin says "some serious eaters think of as their very own Emancipation Proclamation" because a system of quotas excluded Asians and greatly restricted immigration from almost everywhere else from the early 1920s until 1965. But waves of Italian, Chinese, Japanese, and Mexican immigrants during the nineteenth century had provided plenty of opportunities for earlier generations of food adventurers. Trillin, p. 71; Pillsbury, p. 158; Gabaccia, p. 7.

21. Gabaccia, pp. 99–105 (contains Sermolino quote); Barbas.

22. Gabaccia, pp. 231.

23. Barbas.

24. Heldke. For the longer history of the term "social capital," see Robert Putnam, *Bowling Alone* (New York: Simon & Schuster, 2001), pp. 19–20.

25. David Bell, "Fragments for a New Urban Culinary Geography," *Journal for the Study of Food and Society* 6 (2002): 10–21.

26. Samantha Kwan, "Consuming the Other: Ethnic Food, Identity Work, and the Appropriation of the Authentic Self," paper presented at the meeting of the American Sociological Association, August 2003.

27. Sylvia Ferrero, "Comida sin Par: Consumption of Mexican Food in Los Angeles," pp. 194–219 in Warren Belasco and Philip Scranton, eds., *Food Nations* (New York: Routledge, 2002; quotes are from p. 215).

28. See also Ron Howell, "Vegan Restaurants Are Thriving in the Black Community As People Seek a More Healthful Lifestyle," *Newsday,* April 1, 2003.

29. CSPI news release is July 18, 1994; Marian Burros, "A Study Faults Mexican Restaurants," *New York Times,* July 19, 1994; Richard Rodriguez, "Mexican Food," *Los Angeles Times,* July 24, 1994. For more recent examples of CSPI's condemnation of Mexican restaurant food, see, e.g., "Mucho Grosso," *Nutrition Action Newsletter* (December 2001); and "Fresh Mex: Not Always Healthy Mex," CSPI press release, September 30, 2003.

30. Netta Davis, "To Serve the 'Other': Chinese-American Immigrants in the Restaurant Business," *Journal for the Study of Food and Society* 6 (2002): 70–81.

31. Bill Buford, "The Secrets of Excess," *New Yorker*, August 19, 2002, pp. 122–41 (quote is from p. 126).

32. Nicole Mones, "Kitchen Warriors," *Gourmet* (October 2003): 271–75 (quote is from p. 272).

33. Davis; quote is from p. 79; I have converted it to standard from broken English.

34. "Food Courts: Fattening Americans As They Shop," CSPI press release dated March 22, 2001.

35. Barbas; Pillsbury.

36. Alan Davidson, *The Oxford Companion to Food* (New York: Oxford University Press, 1999), p. 182.

37. This thumbnail history is culled from corporate histories from the companies, and Gabaccia; Eric Schlosser, *Fast Food Nation* (Boston: Houghton Mifflin, 2001); and Levenstein, *Paradox of Plenty* (New York: Oxford University Press, 1993).

38. See Levenstein, *Paradox of Plenty*, p. 234, for a different but related interpretation.

Chapter 6: Restaurant Hell

1. Patricia Unterman, "Face to Face with Food," *San Francisco Examiner*, December 3, 2003; Gersh Kuntzman, "Breast Intentions," Newsweek.com, November 24, 2003; Egullet discussion: forums.egullet.com/index.php?s=38874 a1daad394c5bfc7a02a083df6b7&act=ST&f=2&t=28870&st=0.

2. The *Frontera* newsletter is dated Winter 2003.

3. Bayless's letter appeared at www.fronterakitchens.com/rickbayless/what-supwbk/.

4. Phil Vettel, "Why Is Rick Bayless Doing Burger King Commercials," *Chicago Tribune*, October 13, 2003.

5. Bayless quote is from Vettel.

6. Jennifer Parker Talwar, *Fast Food, Fast Track* (Boulder, CO: Westview, 2002), pp. 65–67. See also Robin Leidner, *Fast Food, Fast Talk* (Berkeley: University of California Press, 1993), p. 132.

7. The Charlie X article is posted at www.phrack.org/show.php?p=45&a=19. See also Joe Kincheloe, *The Sign of the Burger* (Philadelphia: Temple University Press, 2002), pp. 36–43.

8. Elspeth Probyn, *Carnal Appetites* (New York: Routledge, 2000), p. 35 (and see also p. 51, and Kincheloe, p. 37).

9. Kincheloe, p. 44. The author is critical of this and other customers' favorable views of McDonald's, for reasons I discuss and assess critically in my review of the book in the journal *Gastronomica* (Summer 2003): 95–96.

10. Frank McNally, "Food Is the New Smoking," *Irish Times,* November 16, 2002.

11. See Frank Rich, "The Mighty Mo," *New York Times,* October 6, 1999.

12. Constantin Boym, "My McDonald's," *Gastronomica* (Winter 2001): 6–8.

13. Jeet Heer and Steve Penfold, "The Resilence of Regional Identity," *Responsive Community* (Summer 2003): 6–11; Rick Fantasia, "Everything and Nothing," *Tocqueville Review* 15, no. 2 (1994): 57–88.

14. Les Gapay, "How a Regular Guy Gets Homeless," *USA Today,* September 22, 2003.

15. The government's "Thrifty Food Plan" is available at the USDA Web site, www.usda.gov. See also Jeffrey Steingarten, *The Man Who Ate Everything* (New York: Random House, 1997), pp. 33–49.

16. See, for example, Jennifer Orlet Fisher and Leann Birch, "Eating in the Absence of Hunger and Overweight in Girls from 5 to 7 Years of Age," *American Journal of Clinical Nutrition* 76, no. 1 (2002): 226–231.

17. John Coveney, *Food, Morals and Meaning* (New York: Routledge, 2000), p. 24.

18. Harvey Levenstein, *Revolution at the Table* (New York: Oxford University Press, 1988), pp. 98–99.

19. Levenstein, *Revolution at the Table,* p. 56. See also Ben Rogers, *Beef and Liberty* (London: Chatto & Windus, 2003).

20. Levenstein, *Revolution at the Table,* p. 21. The cookbook is M. Tarbox Colbrath, *What to Get for Breakfast* (Boston: James H. Earle, 1883).

21. Harvey Levenstein, *Paradox of Plenty,* p. 228 (Ingram quotes); David Gerard Hogan, *Selling 'Em by the Sack* (New York: New York University Press, 1997), chaps. 1–2; "White Castle Corporate History," Ohio Historical Society, MSS 991 (available at www.ohiohistory.org).

22. Hogan, pp. 32–34 and illustration section.

23. Hogan, chap. 3; Eric Schlosser, *Fast Food Nation* (Boston: Houghton Mifflin, 2001); Levenstein, *Paradox of Plenty,* chap. 15.

24. Ray Kroc, *Grinding It Out* (Chicago: Contemporary Books, 1977) (quotes are from pp. 14, 75, and 91).

25. Elaine Sciolino, "Adrià Turns the Charms of El Bulli into Fast Food," *New York Times,* July 28, 2004; Schlosser, *Fast Food Nation* (quotes are from pp. 33 and 221); Eric Schlosser, "How to Make the Country's Most Dangerous Job Safer," *New Yorker* (January 2002): 34–35; Eric Schlosser, "Human Rights Are Dying on the Vine," *Los Angeles Times,* March 5, 2004.

26. Quote is in Schlosser, *Fast Food Nation,* p. 264. See in Schlosser, "How to Make the Country's Most Dangerous Job Safer"; Schlosser, "Human Rights Are Dying on the Vine."

27. Warren Belasco, "Roadside Dreams, Fast Food Nightmares," *Technology and Culture* 42 (2001): 767–70; George Orwell, *Down and Out in Paris and London* (New York: Penguin, 1968), chap. 14; Anthony Bourdain, *Kitchen*

Confidential (New York: Bloomsbury: 2000); Gary Alan Fine, *Kitchens* (Berkeley: University of California Press, 1996).

28. Claudia Deutsch, "The Arches Are Sagging," *New York Times*, October 20, 2002; Bruce Horovitz, "Fast Food Giants Try Value Menus," *USA Today*, September 12, 2002.

29. Daniel Akst, "Cheap Eats," *Wilson Quarterly* 27 (Summer 2003): 30–41; Claire Hope Cummings, "Entertainment Foods," *Ecologist* 29 (1999): 16–19; Schlosser, *Fast Food Nation*, p. 72 and epilogue.

30. Schlosser, *Fast Food Nation*, pp. 47 and 79.

31. See Schlosser, *Fast Food Nation*, on the first list; Cummings; Kincheloe, p. 106.

32. Kincheloe ("hegemonic" reference is from pp. 151–52; quote is from p. 48); George Ritzer, *The McDonaldization of Society* (Thousand Oaks, CA: Pine Forge Press, 2004). See also Nick Perry, "Traveling Theory, Nomadic Theorizing," *Organization* 2 (1995): 35–54; and for some of the more insightful elaborations and critiques of the McDonaldization thesis, see Barry Smart, ed., *Resisting McDonaldization* (Thousand Oaks, CA: Sage, 1999).

33. Schlosser, *Fast Food Nation*; Kincheloe, p. 34.

34. The statistics and arguments supporting them are posted at www.earthworksaction.com.

35. Kroc, p. 199.

36. See Kroc, p. 180.

37. Quoted from www.slowfoodusa.org/about/principles.html.

38. Fernandez-Armesto, *Near a Thousand Tables* (New York: Free Press, 2002), p. 223; Elisabeth Rozin, *The Primal Cheesebuger* (New York: Penguin, 1994; quotes are from pp. 6 and 37). On the primal appeal of meat, and the evolved genetic basis for this, see Caleb Finch and Craig Stanford, "Meat-Adaptive Genes and the Evolution of Slower Aging in Humans," *Quarterly Review of Biology* 79, no. 1 (2004): 3–50.

39. Rozin, pp. 39 and 61.

40. Here I depart slightly from Rozin's analysis (see p. 59), which focuses on handheld food coming closer to the nose.

41. Rozin, p. 83.

42. Rozin, chap. 4; Davidson, *Oxford Companion to Food*, p. 430. See also Andrew Smith, *Pure Ketchup* (Columbia: University of South Carolina Press, 1996).

43. See Hogan.

44. Stacy Perman, "Fat Burgers," *Los Angeles Magazine* (February 2004): 36–41; Tom McNichol, "The Secret Behind a Burger Cult," *New York Times*, August 14, 2002 (contains Schlosser quote); Patrick McGeehan, "The Red Carpet Leads to a Drive-Through," *New York Times*, March 7, 2004.

45. Schlosser, *Fast Food Nation*, chap. 5.

46. Talwar, pp. 58, 70–71. See also Robin Leidner, chap. 3; Schlosser, *Fast Food Nation:* 71–75.

47. Talwar, chap. 8 (contains Hagans's story and similar examples). Motes's story appears on Burger King's Web site, www.bk.com.

48. Kroc, p. 111 (contains quote). Cost information is from Kroc, p. 178; Burger King's and McDonald's Web sites; www.entrepreneur.com; and "McDonald's Makes Franchising Sizzle," *BusinessWeek,* June 15, 1968, pp. 102–3.

49. Talwar, chap. 8.

50. Talwar (quote is from p. 2).

51. Quote is from Talwar, p. 2. See also David Shipler, *The Working Poor* (New York: Knopf, 2004), p. 19.

52. Schlosser, *Fast Food Nation,* pp. 75–83, 265.

53. Recycling quote is from Donna Fenn, "Veggie-Burger Kings," *Inc.* (November 2001): 44.

54. McDonald's quote is from the company's Web site; Anderson is quoted in Josef Woodard, "Her Private Happy Meal," *Los Angeles Times,* January 27, 2002. When one of my graduate students at the University of Southern California, Steve Zafirau, monitored online discussion groups for McDonald's workers and interviewed employees at a McDonald's restaurant where he himself worked, he found a full spectrum of reactions to working under the Golden Arches. At one end are people who find the job unrewarding, demeaning, and exhausting; at the other are those who enjoy working there and consider their fellow employees family. Steve Zafirau, "Branded Identities at Work," unpublished paper, 2004.

55. Talwar, pp. 76–77.

56. David Harris, "Fast Food Moves Beyond the Burger," *Boston Globe,* January 14, 2004; menu information at the O'Naturals Web site, www.onaturals.com.

57. See Sherry Ruth Anderson, *The Cultural Creatives* (New York: Three Rivers Press, 2001).

58. Rich Ganis, "Burger King Uncowed," *Los Angeles Times,* April 14, 2002.

Chapter 7: What Made America Fat?

1. Robert J. Kuczmarski, K. M. Flegal, et al., "Increasing Prevalence of Overweight Among U.S. Adults," *Journal of the American Medical Association* 272 (1994): 205–11; Katherine M. Flegal, M. D. Carroll, et al., "Overweight and Obesity in the United States: Prevalence and Trends, 1960–1994," *International Journal of Obesity and Related Metabolic Disorders* 22 (1998): 39–47.

2. Schlosser, *Fast Food Nation* (Boston: Houghton Mifflin, 2001), p. 240. Obesity statistics for 1960–2000 are from the National Center for Health Statistics publication, *Health, United States, 2003,* pp. 230–33. Fast-food time lines are from Kroc, *Grinding It Out* (Chicago: Contemporary Books, 1977), and corporate histories posted on McDonald's and Burger King's Web sites. In the 1980s, the fast-food industry continued to grow, but as Gary Taubes has pointed out, consumption of its products "did not take a sudden leap, as obe-

sity did" (Taubes, "What If It's All Been a Big Fat Lie?" *New York Times*, September 10, 2002). The per capita number of restaurants of all types increased substantially during the 1980s and 1990s, but as we will see, only some segments get blamed for obesity. Moreover, rapid growth in the number of restaurants began at least a decade before obesity rates took off. See figures 1 and 2 in Inas Rashad, Michael Grossman, and Shin-Yi Choua, "The Super Size of America," unpublished paper, November 2003. A subsequent paper by two of these authors includes a caution against implicating restaurants as a cause of the obesity epidemic. "A very different interpretation emerges if one recognizes that the growth in these restaurants, and especially fast-food restaurants, is to a large extent a response to the increasing scarcity and increasing value of household or non-market time," they write in Shin-Yi Choua, Michael Grossman, and Henry Saffer, "An Economic Analysis of Adult Obesity," *Journal of Health Economics* 23 (2004): 565–87.

3. S. Bryn Austin, "Fat, Loathing, and Public Health," *Culture, Medicine and Psychiatry* 23 (1999): 245–68. Austin credits the term "fiscal model" to William Bennett, "Dieting: Ideology Versus Physiology," *Psychiatric Clinics of North America* 7 (1984): 321–34. Quote is from Amanda Spake, "Rethinking Weight," *U.S. News & World Report*, February 9, 2004. Other examples: Tara Parker-Pope, "If You Want to Lose Weight, Cut Down on Calories," *Wall Street Journal*, January 21, 2003; Nanci Hellmich, "Experts: Eat Less, Exercise More. Any Questions," *USA Today*, July 19, 2004. Almost invariably, even in discussions of the relative merits of one sort of food over another, the bottom line in reports by medical and governmental panels and health writers has been that "to maintain weight the number of calories consumed should not exceed the number of calories expended" (Marian Burros, "Panel Advises More Focus on Grains, Less on Sugar," *New York Times*, August 11, 2004).

4. Gina Kolata, "While Children Grow Fatter, Experts Search for Solutions," *New York Times*, October 19, 2000.

5. Peters quote is from Joan Brumberg, *Fasting Girls* (New York: Penguin, 2000), p. 239, and is included in Austin; and see Natalie Boero, "All the News That's Fat to Print," paper presented at the meeting of the American Sociological Association, August 2003.

6. William Bennett and Joel Gurin, *The Dieter's Dilemma* (New York: Basic Books, 1982; quote is from p. xiii); Bennett, "Dieting: Ideology Versus Physiology"; William Bennett, "Beyond Overeating," *New England Journal of Medicine* 332 (1995): 673–74. For a related critique, see Theodore Van Itallie, "Dietary Approaches to the Treatment of Obesity," pp. 249–61 in Albert J. Stunkard, ed., *Dietary Approaches to the Treatment of Obesity* (Philadelphia: W. B. Saunders, 1980).

7. Bennett, citations above (quotes are from pp. 60 and 17 in Bennett and Gurin); Paul Thomas and Judith Stern, *Weighing the Options: Criteria for Evaluating Weight-Management Programs* (Washington, D.C.: National Academy

Press, 1995); David Garner and Susan Wooley, "Confronting the Failure of Behavioral and Dietary Treatments for Obesity," *Clinical Psychology Review* 11 (1991): 729–80; "Calorie Intake Same for Larger Persons," *Healthy Weight Journal* 13 (1999): 2; Claude Bouchard and Angelo Tremblay, "Genetic Influences on the Response of Body Fat and Fat Distribution to Positive and Negative Energy Balances in Human Identical Twins," *Journal of Nutrition* 127 (1997): 943–47. In personal correspondence, scientists at the U.S. Department of Agriculture and the Centers for Disease Control who specialize in diet, nutrition, and weight confirmed that studies commonly find an inverse relationship between BMI and food energy intake. Some pointed to consumer research as well. ("My favorite example: Practically all obese people consume diet soft drinks as opposed to regular soft drinks," wrote one.) These experts underscored, however, a caveat I discussed earlier regarding studies of food consumption, namely, the likelihood of inaccuracies in people's reports of what they eat. There is the possibility that obese men and women underreport their calorie intake. Some experiments also suggest that heavier people eat more; see for example Cara B. Ebbeling, Kelly B. Sinclair, et al., "Compensation for Energy Intake from Fast Food Among Overweight and Lean Adolescents," *Journal of the American Medical Association* 291 (2004): 2828–33.

8. Robert Pool, *Fat: Fighting the Obesity Epidemic* (New York: Oxford University Press, 2001), p. 153 ("food-rich" quote); Critser, *Fat Land* (New York: Houghton Mifflin, 2003), p. 53.

9. Jeffrey M. Friedman, "A War on Obesity, Not the Obese," *Science* 299 (2003): 856–58. See also Gregory S. Barsh, I. S. Farooqi, and S. O'Rahilly, "Genetics of Body-Weight Regulation," *Nature* 404 (2000): 644–51; Karolin Schousboe: M. Visscher, et al., "Twin Study of Genetic and Environmental Influences on Adult Body Size, Shape and Composition," *International Journal of Obesity* 28 (2004): 39–48; L. Qi, H. Larson, et al., "Gender-Specific Association of Perilipin Gene Haplotype with Obesity Risk in a White Population," *Obesity Research* 12 (2004): 1758–65; Jeffrey M. Friedman, "Modern Science Versus the Stigma of Obesity," *Nature Medicine* 10 (June 2004): 563–69.

10. For these and related statistics and first-person accounts, see David Shipler, *The Working Poor* (New York: Knopf, 2004); Katherine Newman, *Falling from Grace* (Berkeley: University of California Press, 1999); "Income Inequality," report published by Americans for Democratic Action, Washington D.C., 2004; William Welch, "Report: 82M Went Uninsured," *USA Today,* June 16, 2004; "Can We Give America a Raise," *American Prospect* (January 2004); Robert Perrucci and Earl Wyson, *The New Class Society* (New York: Rowman & Littlefield, 2003).

11. On SES, mobility, and obesity, see, for example, Albert J. Stunkard, "Socioeconomic Status and Obesity," pp. 174–206 in Derek Chadwick and Gail Cardew, eds., *The Origins and Consequences of Obesity* (New York: Wiley, 1996); U.S. Department of Health and Human Services, *Healthy People 2010* (Wash-

ington, D.C.: U.S. Government Printing Office, 2000); Claudia Langenberg, R. Hardy, et al., "Central and Total Obesity in Middle Aged Men and Women in Relation to Lifetime Socioeconomic Status," *Journal of Epidemiology and Community Health* 57 (2003): 816–22. The inverse relationship between SES and weight is stronger among women than men: Katherine M. Flegal, M. D. Carroll, et al., "Prevalence and Trends in Obesity Among US Adults 1999–2000," *Journal of the American Medical Association* 288 (2002): 1723–27.

12. Mary Dallman, Norman Pecoraro, et al., "Chronic Stress and Obesity," *Proceedings of the National Academy of Sciences* 100 (2003): 11696–701; Elissa Epel, R. Lapidus, et al., "Stress May Add Bite to Appetite in Women," *Psychoneuroendocrinology* 26 (2001): 37–49; Elissa Epel, B. McEwen, et al., "Stress and Body Shape," *Psychosomatic Medicine* 62 (2000): 623–32; Roland Rosmond and Per Björntorp, "Occupational Status, Cortisol Secretory Pattern, and Visceral Obesity in Middle-Aged Men," *Obesity Research* 8 (2000): 445–50; Rajita Sinha, M. Talih, et al., "Hypothalamic-Pituitary-Adrenal Axis and Sympatho-Adreno-Medullary Responses During Stress-Induced and Drug Cue–Induced Cocaine Craving States," *Psychopharmacology* 170 (2003): 62–72; Elissa Epel, Nancy Adler, et al., "Nocturnal Growth Hormone Excretion in Healthy Women: Relationships to Acute Stress, Psychological Stress-Resilience, and Abdominal Fat Distribution," unpublished paper, 2004.

13. On the effects of antismoking campaigns on the obesity epidemic, see also Daniel Gross, "Cigarettes, Taxes and Thin French Women," *New York Times,* July 25, 2005; and Choua et al., which provides evidence for an economic link in addition to the physiological one I have noted here.

14. Robert K. Merton, "The Unanticipated Consequences of Purposive Social Action," *American Sociological Review* 1 (1936): 894–904. See also Jan Elster, "Merton's Functionalism and the Unintended Consequences of Action," pp. 128–35 in Jon Clark et al., eds., *Robert K. Merton: Consensus and Controversy* (New York: Falmer Press, 1990).

15. Lena Williams, "Growing Up Flabby in America," *New York Times,* March 22, 1990; Boero, p. 34.

16. I could find only one study (Patricia Anderson et al., "Maternal Employment and Overweight Children," *Journal of Health Economics* 22 [2003]: 477–504) that reported a direct and statistically significant relationship between working mothers and childhood obesity, and only with major caveats. The researchers concluded that maternal employment accounts for just a small portion of the rise in childhood obesity, and principally within higher-income families where mothers work long hours. They note as well that another study (Rachel K. Johnson et al., "Maternal Employment and the Quality of Young Children's Diets," *Pediatrics* 90 [1992]: 245–49) found no relationship. See also Helen Sweeting and P. West, "Dietary Habits and Children's Family Lives," *Journal of Human Nutrition & Dietetics* 18 (2005): 93–97. On time with parents, see for example John Sandberg and Sandra Hofferth, "Changes in Children's Time with Parents," *De-*

mography 38 (2001): 423–36; Stephanie Coontz, *The Way We Never Were* (New York: Basic Books, 2000).

17. Undurti Das, "Is Obesity an Inflammatory Condition?" *Nutrition* 17 (2001): 974–75. On trends in breast-feeding versus bottle-feeding, see Marilyn Yalom, *A History of the Breast* (New York: Knopf, 1997), chap. 4; Pam Carter, *Feminism, Breasts, and Breast-Feeding* (New York: St. Martin's, 1995), chap. 2; Clifford Lo and Ronald Kleinman, "Infant Formula, Past and Future," *American Journal of Clinical Nutrition* 63 (1996): 646. Also see the set of papers on breast-feeding and obesity in the October 18, 2003, issue of the *British Journal of Medicine.*

18. Gregg Easterbrook, "Wages of Wealth: All This Progress Is Killing Us, Bite by Bite," *New York Times,* March 14, 2004.

19. Kuntzman, "Breast Intentions," Newsweek.com, November 24, 2003; Marie-Pierre St.-Onge, Kathleen Keller, and Steven Heymsfield, "Changes in Childhood Food Consumption Patterns," *American Journal of Clinical Nutrition* 78 (2003): 1068–73.

20. Ebbeling, Sinclair, et al.; Cynthia Ogden, Katherine M. Flegal, et al., "Prevalence and Trends in Overweight Among US Children and Adolescents, 1999–2000," *Journal of the American Medical Association* 288 (2002): 1728–32.

21. Critser, p. 115; Simone A. French, M. Story, et al., "Fast Food Restaurant Use Among Adolescents," *International Journal of Obesity* 25 (2001): 1823–33.

22. Todd G. Buchholz, "Burgers, Fries, and Lawyers," *Policy Review Online* (March 2004).

23. The "summit" was held in Williamsburg, Virginia, June 2–4, 2004.

24. On Arkansas food, see John T. Edge, *Southern Belly* (Athens, GA: Hill Street Press, 2000), pp. 27–45.

25. Alison E. Field, S. B. Austin, et al., "Relation Between Dieting and Weight Change Among Preadolescents and Adolescents," *Pediatrics* 112 (2003): 900–906. See also Richard S. Strauss, "Self-Reported Weight Status and Dieting in a Cross-Sectional Sample of Young Adolescents," *Archives of Pediatric and Adolescent Medicine* 153 (1999): 741–47; David F. Williamson, Mary K. Serdula, et al., "Weight Loss Attempts in Adults," *American Journal of Public Health* 82 (1992): 1251–57; Maarit Korkeila, Aila Rissanen, et al., "Weight-Loss Attempts and Risk of Major Weight Gain," *American Journal of Clinical Nutrition* 70 (1999): 965–75.

26. Jayne A. Fulkerson, Megan T. McGuire, et al., "Weight-Related Attitudes and Behaviors of Adolescent Boys and Girls Who Are Encouraged to Diet by Their Mothers," *International Journal of Obesity* 26 (2002): 1579–87.

27. See Jennifer Orlet Fisher and Leann L. Birch, "Eating in the Absence of Hunger and Overweight in Girls from 5 to 7 Years of Age," *American Journal of Clinical Nutrition* 76 (2002): 226–31, and studies cited there.

28. Joanne P. Ikeda, Patricia Lyons, et al., "Self-Reported Dieting Experiences of Women with Body Mass Indexes of 30 or More," *Journal of the Ameri-*

can Dietetic Association 104 (2004): 972–75. See also Linda Bacon et al., "Size Acceptance and Intuitive Eating Improve Health for Obese, Female Chronic Dieters," *Journal of the American Dietetic Association* 105 (2005): 929–36.

29. Quotes are from an interview with Ikeda on Public Radio International's *To the Point,* December 30, 2003.

30. Laura Fraser, *Losing It* (New York: Plume, 1998), p. 122. For additional studies suggesting this conclusion, see Fisher and Birch; and Janet Polivy, "Psychological Consequences of Food Restriction," *Journal of the American Dietetic Association* 96 (1996): 589–92.

31. Religion: Kenneth F. Ferraro, "Firm Believers? Religion, Body Weight, and Well-Being," *Review of Religious Research* 39 (1998): 224–44; Kate Lapane, T. Lasater, et al., "Religious Affiliation and Cardiovascular Disease Risk Factor Status," *Journal of Religion and Health* 36 (1997): 155–63; K. H. Kim, Jeffrey Sobal, and Elaine Wethington, "Religion and Body Weight," *International Journal of Obesity* 27 (2003): 469–77. See the last of these for caveats and studies at variance with these conclusions.

32. Group meals: John M. de Castro and Elaine Brewer, "The Amount Eaten in Meals by Humans Is a Power Function of the Number of People Present," *Physiology and Behavior* 51 (1992): 121–25. Some researchers who implicate restaurant meals in the obesity epidemic cite this phenomenon as a possible cause. See, for example, Megan A. McCrory, Vivian Suen, and Susan B. Roberts, "Biobehavioral Influences on Energy Intake and Adult Weight Gain," *Journal of Nutrition* 132 (2002): 3830–34. Evidence from more recent research by the psychologist who discovered the phenomenon contradicts that supposition, however. See Sharon M. Pearcey and John M. de Castro, "Food Intake and Meal Patterns of Weight-Stable and Weight-Gaining Persons," *American Journal of Clinical Nutrition* 76 (2002): 107–12.

33. Sharada D. Vangipuram et al., "A Human Adenovirus Enhances Preadipocyte Differentiation," *Obesity Research* 12 (2004): 770–77; National Public Radio, *Talk of the Nation,* May 10, 2002; Matt Crenson, "Experiments Suggest 'Fat Virus,'" Associated Press, July 27, 2000; Richard L. Atkinson, Nikhil V. Dhurandhar, et al., "Human Adenovirus-36 Is Associated with Increased Body Weight and Paradoxical Reduction of Serum Lipids," *International Journal of Obesity* 29 (2005): 281–86; Mitali Kapila, Pramod Khosla, and Nikhil V. Dhurandhar, "Novel Short-Term Effects of Adenovirus Ad-36 on Hamster Lipoproteins," *International Journal of Obesity* 28 (2004): 1521–27.

34. David J. Gunnel, M. Okasha, et al., "Height, Leg Length, and Cancer Risk," *Epidemiologic Reviews* 23 (2001): 313–42; Bianca L. De Stavola, I. dos Santos Silva, et al., "Childhood Growth and Breast Cancer," *American Journal of Epidemiology* 159 (2004): 671–82; Klaus P. Dieckmann and U. Pichlmeier, "Is Risk of Testicular Cancer Related to Body Size?" *European Urology* 42 (2002): 564–69; Gunnar Andersson, "Epidemiological Features of Chronic Low-Back Pain," *Lancet* 354 (1999): 581–85; Malcolm I. V. Jayson, "Back

Pain," *British Medical Journal* 313 (1996): 355–59; Charles Slemenda, "Skeletal Factors Other Than BMD That May Increase the Risk of Hip Fracture in Women Include Hip Geometry and Height," *American Journal of Medicine* 103 (1997): 65S–71S; Edith M. Lau: Eggar, et al., "Low Back Pain in Hong Kong, Prevalence and Characteristics Compared with Britain," *Journal of Epidemiology and Community Health* 49 (1995): 492–94; Diana J. Kuh and D. Coggan, "Height, Occupation and Back Pain in a National Prospective Study," *British Journal of Rheumatology* 32 (1993): 911–16; Yin B. Yip, S. C. Ho, and S. G. Chan, "Tall Stature, Overweight and the Prevalence of Low Back Pain in Chinese Middle-Aged Women," *International Journal of Obesity* 25 (2001): 887–92; Xuemei Luo, Ricardo Pietrobon, et al., "Estimates and Patterns of Direct Health Care Expenditures Among Individuals with Back Pain in the United States," *Spine* 29 (2004): 79–86. As various of these studies suggest, the costs of tallness are considerable. Note, for instance, that taller women more often suffer hip fractures, which result in expensive surgeries and other treatments, and taller people have higher rates of chronic back pain, an infirmity that results in $90 billion annually in health-care expenses and is the most common reason people younger than forty-five have to limit their activities and miss work. Paralleling an ambiguity I discuss below regarding the nature of the association between obesity and disease, however, some researchers suggest tallness is rightly understood as a cause of the adverse outcomes, while others argue that both tallness and the adverse health outcomes with which it is associated result from other processes (abnormal levels of growth hormones in childhood, for example).

35. Nanci Hellmich, "Obesity on Track as Number One Killer," *USA Today*, May 10, 2004.

36. Eliot Marshall, "Public Enemy Number One: Tobacco or Obesity?" *Science* 7 304 (May 2004): 804.

37. For a technical critique of the *JAMA* paper, see Katherine M. Flegal, B. Graubard, et al., "Methods of Calculating Deaths Attributable to Obesity," *American Journal of Epidemiology* 160 (2004): 331–38.

38. Kevin R. Fontaine, David T. Redden, et al, "Years of Life Lost Due to Obesity," *Journal of the American Medical Association* 289 (2003): 187–93.

39. For some studies that support this conclusion, see Paul Campos, *The Obesity Myth* (New York: Gotham, 2004), pp. 10–20. I cite other studies on this matter below in other contexts.

40. Katherine Flegal, Barry Graubard, et al., "Excess Deaths Associated with Underweight, Overweight, and Obesity," *Journal of the American Medical Association* 293 (2005): 1861–67. See also Gina Kolata, "Some Extra Heft May Be Helpful, New Study Says," *New York Times*, April 20, 2005. On the methodological superiority of this study to previous studies, see Gina Kolata, "Study Tying Longer Life to Extra Pounds Draws Fire," *New York Times*, May 27, 2005. True to form, Willett blasted the study, and reporters dutifully cited his mistaken assumption that the authors had failed to take into account the effects of prior

illnesses, smoking, and old age within their sample. (See, for example, Rosie Mestel, "Weight Death Toll Downplayed," *Los Angeles Times*, April 20, 2005.) In reality, when Flegal et al. ran further analyses on their massive data set to test Willett's alternative hypotheses, they discovered not only that he was wrong, but that in some cases the findings were in the opposite direction of his contentions. (See "Underweight, Overweight, Obesity, and Excess Death," *Journal of the American Medical Association* 294 [2005]: 551–53.)

41. By highly obese, these authors typically mean a BMI over about 37.

42. Earlier books: Bennett and Gurin; Fraser; Janet Polivy, *Breaking the Diet Habit* (New York: Basic, 1985); Hillel Schwartz, *Never Satisfied* (New York: Simon & Schuster, 1986); Barry Glassner, *Bodies* (New York: Putnam, 1988); Susan Bordo, *Unbearable Weight* (Berkeley: University of California Press, 1993); Glenn A. Gaesser, *Big Fat Lies* (Carlsbad, CA: Gurze Books, 2002).

43. Campos, p. 21; Nanci Hellmich, "Everyone Knows by Now That Fat Kills You. Or Does It?" *USA Today*, May 4, 2004.

44. Gaesser, chap. 3.

45. Walter Willett, JoAnn Manson, et al., "Weight, Weight Change, and Coronary Heart Disease in Women," *Journal of the American Medical Association* 273 (1995): 461–65.

46. Susanna Jonsson, B. Hedblad, et al., "Influence of Obesity on Cardiovascular Risk," *International Journal of Obesity* 26 (2002): 1046–53.

47. JoAnn Manson, Walter Willett, et al., "Body Weight and Mortality Among Women," *New England Journal of Medicine* 333 (1995): 677–85. The numbers appear atop figure 3. (Figure 4 looks at all deaths in the sample from cardiovascular disease—716 women—but the findings do not make Willett et al.'s point so neatly. Women in the 29–32 BMI range have lower death rates than those in the 27–29 range.) My analysis here elaborates on Fraser, pp. 12–14. For incisive commentaries about statistical ambiguities in these papers by Willett and his colleagues, see Austin.

48. Susan Z. Yanovski and Jack A. Yanovski, "Obesity," *New England Journal of Medicine* 346 (2002): 591–602. See also Gina Kolata, "Diet and Lose Weight? Scientists Say 'Prove It,' " *New York Times*, January 4, 2005.

49. The sole study of weight loss to which Willett referred me does not support his contention about weight loss. In this study of 101 overweight people put on either a moderate-fat or low-fat diet for eighteen months, about 40 percent of the subjects dropped out entirely, and a grand total of thirty-seven were still on the diets at the end of the eighteen months. Willett and the paper's authors both emphasized that nineteen people in a subsample were almost eight pounds below their initial weight thirty months after the start of the study. But even that favorable finding is offset by the fact that twenty-six of those who dropped out of the study and allowed themselves to be weighed after eighteen months had an average weight gain of nine pounds. Katherine McManus, Linda Antinoro, and Frank Sacks, "A Randomized Controlled Trial

of a Moderate-Fat, Low-Energy Diet Compared with a Low-Fat, Low-Energy Diet for Weight Loss in Overweight Adults," *International Journal of Obesity* 25 (2001): 1503–11.

50. Richard Klein, "Big Country," *New Republic*, September 19, 1994, pp. 28–33; Sander Gilman, *Fat Boys* (Lincoln: University of Nebraska Press, 2004); Peter N. Stearns, *Fat History* (New York: New York University Press, 2002); Glassner, *Bodies*; Eric Oliver, *Obesity: The Making of an American Epidemic* (New York: Oxford University Press, 2005).

51. Thorkild I. A. Sørensen, "Weight Loss Causes Increased Mortality: Pros," *Obesity Reviews* 4 (2003): 3–7; Sven Rossner, "Does Sustained Weight Loss Lead to Decreased Morbidity and Mortality?—No," *International Journal of Obesity* 23, suppl. 5 (1999): S41; Edward W. Gregg and David F. Williamson, "Relationship of Intentional Weight Loss to Disease Incidence and Mortality," in T. Wadden and A. Stunkard, eds., *Handbook of Obesity Treatment* (New York: Guilford Press, 2002); Gaesser, chap. 7; Linda Nebeling, Connie J. Rogers, et al., "Weight Cycling and Immunocompetence," *Journal of the American Dietetic Association* 104 (2004): 892–94. For people with certain preexisting chronic conditions, intentional weight loss does appear to reduce mortality. See, for example, Edward W. Gregg, Theodore J. Thompson, et al., "Trying to Lose Weight, Losing Weight, and 9-Year Mortality in Overweight U.S. Adults with Diabetes," *Diabetes Care* 27 (2004): 657–62.

52. Jerome Kassirer and Marcia Angell, "Losing Weight—An Ill-Fated New Year's Resolution," *New England Journal of Medicine* 338 (1998): 52–54.

53. On set point theory, see also Barry E. Levin and Richard E. Keesey, "Defense of Differing Body Weight Set Points in Diet-Induced Obese and Resistant Rats," *American Journal of Regulatory, Integrative and Comparative Physiology* 274 (1998): R412–R419; Richard E. Keesey and Matt D. Hirvonen, "Body Weight Set-Points: Determination and Adjustment," *Journal of Nutrition* 127 (1997): 1875S–1883S; Bennett and Gurin; William Bennett, "Beyond Overeating," *New England Journal of Medicine* 332 (1995): 673–74.

54. Mark Roehling, "Weight-Based Discrimination in Employment," *Personnel Psychology* 52 (1999): 969–1016; Steven L. Gortmaker, Aviva Must, et al., "Social and Economic Consequences of Overweight in Adolescence and Young Adulthood," *New England Journal of Medicine* 329 (1993): 1008–12; Janet D. Latner and Albert J. Stunkard, "Getting Worse: The Stigmatization of Obese Children," *Obesity Research* 11 (2003): 452–56; Cathy S. Reto, "Psychological Aspects of Delivering Nursing Care to the Geriatric Patient," *Critical Care Nursing Quarterly* 26 (2003): 139–50; Deborah Carr, "Obesity and Perceived Discrimination in the United States," paper presented at the meetings of the American Sociological Association, 2004; Stephanie Armour, "Your Appearance, Good or Bad, Can Affect Size of Your Paycheck," *USA Today*, July 20, 2005. On the relative importance of discrimination and related factors as compared with "health risk behaviors" such as overweight and obesity, see Paula M. Lantz, John W.

Lynch, et al., "Socioeconomic Disparities in Health Change in a Longitudinal Study of US Adults," *Social Science and Medicine* 53 (2001): 29–40; Paula M. Lantz, James S. House, et al., "Socioeconomic Factors, Health Behaviors, and Mortality," *Journal of the American Medical Association* 279 (1998): 1703–8; James S. Jackson, David R. Williams, and Myriam Torres, "Discrimination, Health and Mental Health: The Social Stress Process," chap. 8 in *Socioeconomic Conditions, Stress and Mental Disorders,* published online by the Mental Health Statistics Improvement Program, 2003; Emilie Agardh, Anders Ahlbom, et al., "Explanations of Socioeconomic Differences in Excess Risk of Type 2 Diabetes in Swedish Men and Women," *Diabetes Care* 27 (2004): 716–21.

55. Janet D. Latner and Albert J. Stunkard; John Lynch, "Income Inequality and Health," *Social Sciences & Medicine* 51 (2000): 1001–5; Eric Brunner, "Stress and the Biology of Inequality," *British Medical Journal* 314 (1997): 1472–76; Richard S. Strauss and Harold A. Pollack, "Social Marginalization of Overweight Children," *Archives of Pediatric and Adolescent Medicine* 157 (2003): 746–52; Annika Rosengren, Lars Wilhelmsen, and Kristina Orth-Gomér, "Coronary Disease in Relation to Social Support and Social Class in Swedish Men," *European Heart Journal* 25 (2004): 56–63. See also Michael Marmot, *Status Syndrome.* At the opposite end of the economic ladder, a cycle of a different sort operates, an auspicious cycle. This cycle, too, is self-sustaining. Shielded from discrimination and chronic stress, and possessed of more social connections and money for higher education, weight-loss drugs, and personal trainers, the rich stay healthier and thinner, and thus better paid, less subject to discrimination, and less stressed.

56. Critser, pp. 174–75. Campos, pp. 234–35, makes a point similar to mine regarding consumption by the wealthy. Jennifer C. Sabel and J. Van Eenwyk, "Food Insecurity as a Risk Factor for Obesity," paper presented at the American Society for Parenteral and Enteral Nutrition's "Nutrition Week" conference, 2002; Marilyn Townsend, J. Peerson, et al., "Food Insecurity Is Positively Related to Overweight in Women," *Journal of Nutrition* 131 (2001): 1738–45; Peter Basiotis and Mark Lino, "Food Insufficiency and Prevalence of Overweight Among Adult Women," *Nutrition Insights* 26 (2002): 1–2; Elizabeth J. Adams, Laurence Grummer-Strawn, and Gilberto Chavez, "Food Insecurity Is Associated with Increased Risk of Obesity in California Women," *Journal of Nutrition* 133 (2003): 1070–74. Some studies define food insecurity not in terms of concerns about having enough to eat but as limited or uncertain availability of nutritionally acceptable or safe foods. See Steven J. Carlson, M. S. Andrews, and G. W. Bickel, "Measuring Food Insecurity and Hunger in the United States," *Journal of Nutrition* 129 (1999): 510–16.

57. Sabel and J. Van Eenwyk.

Chapter 8: Conclusion

1. This estimate is based on figures from the U.S. Department of Agriculture and reports by the largest hunger-relief organizations for the numbers of clients they serve.

2. Robert Egger, *Begging for Change* (New York: HarperBusiness, 2004), p. 85

3. Jeffrey Steingarten, "Perfect Pizza," pp. 22–29 in Ed Levine, *Pizza: A Slice of Heaven* (New York: Universe, 2005); Tony Perry, "Paul Saltman: UC San Diego Biochemist, Innovator," *Los Angeles Times*, August 28, 1999; Sean Henahan, interview with Saltman, posted at www.accessexcellence.org.

4. Hasia Diner, *Hungering for America* (Cambridge, MA: Harvard University Press, 2001), chap. 4.

5. Mark Warbis, "Suit Says Albertson's Forces Unpaid Work," Associated Press, April 22, 1997; "Supermarket Strike Averted," *East Bay Business Times*, January 24, 2005.

6. See also Egger, pp. 107–8; "A Profile of the Working Poor," U.S. Bureau of Labor Statistics, March 2005.

7. Jianghong Liu, Adrian Raine, et al., "Malnutrition at Age 3 Years and Externalizing Behavior Problems at Ages 8, 11, and 17 Years," *American Journal of Psychiatry* 161 (2004): 2005–13; David Shipler, *The Working Poor* (New York: Knopf, 2004), chap. 8; Irwin H. Rosenberg et al., "Statement on the Link Between Nutrition and Cognitive Development in Children," Center on Hunger and Poverty, Brandeis University, 1998.

8. Roy Rivenburg, "Scaling Food Pyramid Makes One Guinea Pig a Lesser Man," *Los Angeles Times*, February 19, 2005.

9. Marion Nestle, *Food Politics* (Berkeley: University of California Press, 2002), pp. 71 and 73. Some of the testimony by industry and advocacy groups at the most recent *Guidelines* hearings is available at www.health.gov/dietary guidelines/dga2005/comments. On the PCRM, see, for example, Mary Carmichael, "Atkins Under Attack," *Newsweek*, February 23, 2004; Joe Sharkey, "Perennial Foes Meet Again in a Battle of the Snack Bar," *New York Times*, November 23, 2004; "Who's Who in Animal Rights," *Observer* (London), August 1, 2004.

10. See, for example, Nestle, *Food Politics*, part 4; "Vitamin A: 'Magic Bullet' That Can Backfire," *Tufts Nutrition Letter* (February 2005); Edgar Miller, Roberto Pastor-Barriuso, et al., "Meta-Analysis: High-Dosage Vitamin E Supplementation May Increase All-Cause Mortality," *Annals of Internal Medicine* 142 (2005): 37–46.

11. USDA Food Guidance System Public Comment Meeting, August 19, 2004: 101 (www.usda.gov/cnpp/pyramid-update/Comments/Oral%20Comm entsTranscript.pdf); Warren Belasco, "Futures Notes: The Meal-in-a-Pill," *Food and Foodways* 8 (2000): 253–71. Except as otherwise noted, my analysis and quoted material below utilize the Belasco article.

12. Quotes are from materials from the California Olive Industry and the Olive Oil Source.

13. See also "General Mills Touts Sugary Cereal as Healthy Kids Breakfast," *Wall Street Journal,* June 22, 2005.

14. Almond ad appeared in *Stagnito's New Products Magazine,* (May 2004).

15. Testimony at the Food Guidance System meeting (see n. 11, p. 294).

16. Keith Seiz, "Promoting Grain Based Foods," *Baking Management* (August 2004).

17. Lisa M. Krieger and Paul Jacobs, "Stem Cell Panelists Show Holdings," *San Jose Mercury News,* January 19, 2005; Benjamin Rosenthal, Michael Jacobson, and Marcy Bohm, "Professors on the Take," *Progressive* (November 1976): 42–47.

18. Paul Griffo, "The Great VNR Debate," *Public Relations Tactics* (June 2004); "Shaping the News: The Public Relations Industry and Journalism," available at www.nationalradioproject.org; and reports at www.prwatch.org. See also David Barstow and Robin Stein, "Under Bush, a New Age of Prepackaged News," *New York Times,* March 13, 2005; Associated Press, "Schwarzenegger Video News Release Gets Bad Reviews," March 1, 2005.

19. Associated Press, "Genetically Modified Food Items Are Common, but Little Noticed," March 24, 2005; "Genetically Modified Corn on the Rise," *Pacific Business News,* September 7, 2004; Kathleen Hart, *Eating in the Dark* (New York: Pantheon, 2002), chap. 1; Lizette Alvarez, "Consumers in Europe Resist Gene-Altered Foods," *New York Times,* February 11, 2003; William K. Hallman, W. Carl Hebden, et al., "Americans and GM Foods: Knowledge, Opinion and Interest in 2004," New Brunswick, NJ, Food Policy Institute, Rutgers University, 2004; "EU Lifts Ban on Genetically-Modified Foods," *Cal Trade Report,* May 20, 2004; "Worried Consumers 'Shun GM Foods,'" BBC News, September 2, 2004.

20. Richard Lewontin, "Genes in the Food," *New York Review of Books,* June 21, 2001; Nina Fedoroff, *Mendel in the Kitchen* (Washington, D.C.: National Academies Press, 2004); Daniel Charles, *Lords of the Harvest* (New York: Perseus, 2001), pp. 303–14; Institute for Food Safety and Technology, "Statement of Genetic Modification and Food," July 2004 (available at its Web site); Associated Press, "Genetically Modified. . . ."

21. Hart, p. 125.

22. David Toke, *The Politics of GM Food* (London: Routledge, 2004), pp. 62–64. Toke goes on to argue, convincingly, for the importance of other differences as well between the U.S. and UK, including greater influence of environmental groups and the tabloid press in Britain, and differences in how risk is assessed and managed in the two nations. His study raises the possibility, as well, that Britons may have been predisposed to respond more negatively to both the mad cow scare and GM food than Americans, for cultural or other reasons.

23. Carlo Petrini, *Slow Food* (New York: Columbia University Press, 2001), p. 8.

24. See, for example, John Antle, "The New Economics of Agriculture," a paper presented at the annual meeting of the American Agricultural Economics Association, August 1999; Richard McGill Murphy, "Truth or Scare," *American Demographics* 26 (March 2004): 26–33.

25. Julie Guthman, "The Trouble with 'Organic Lite' in California," *Sociologia Ruralis* 44 (2004): 301–16 (quote is from p. 305).

26. See, for example, Julie Guthman, *Agrarian Dreams* (Berkeley: University of California Press, 2004); Charles Thompson, "Layers of Loss," pp. 55–86 in Charles Thompson and Melinda Wiggins, eds., *The Human Cost of Food* (Austin: University of Texas Press, 2002).

27. Bruce Gardner, "The Little Guys Are O.K.," *New York Times*, March 7, 2005.

28. Mimi Knight, "The Family That Eats Together," *Christianity Today* (February 2002).

29. Jay Walljasper, "The Joy of Eating," *Utne Reader* (June 2002). See also Probyn, p. 37 (Esko quote).

30. John Gillis, "Making Time for Family," *Journal of Family History* 21 (1996): 4–22.

31. Karlyn Bowman, "The Family Dinner, Alive and Well," *New York Times*, August 25, 1999; Candy Sagon, "Working Parents, Busy Kids, Hectic Lives, but Everything Stops for the Family Meal," *Washington Post*, March 3, 1999; Jodi Spicer, "Who's Coming to Dinner," Family and Consumer Sciences Quarterly Media Pack (Michigan State University), 2003.

32. Dianne Neumark-Sztainer, Melanie Wall, et al., "Are Family Meal Patterns Associated with Disordered Eating Behaviors Among Adolescents?" *Journal of Adolescent Health* 35 (2004): 350–59. Another example: in 1999, journalists solemnly reported that family dinners have a "deterrent effect on drug use by youngsters," though the study that reached this conclusion neglected to consider whether drug use was the chicken or the egg. Conducted by a conservative think tank that decries the decline of family dining, the study may simply have documented a commonplace: when teens start experimenting with drugs, they stop spending as much time with their folks. For an example of coverage of the study, which was released by the National Center on Addiction and Substance Abuse as the "CASA 2000 Teen Survey," see Sarah Fritschner, "Study Shows That Family Dinners Deter Substance Abuse," *Louisville Courier Journal*, September 8, 1999.

33. Elspeth Probyn, *Carnal Appetites* (New York: Routledge, 2000), p. 38 (contains Giard quote); Darra Goldstein, "Food from the Heart," *Gastronomica* (Winter 2004): iii.

34. Steven Shapin, "The Great Neurotic Act," *London Review of Books*, August 5, 2004.

35. Richard Stein and Carol Nemeroff, "Moral Overtones of Food," *Personality and Social Psychology Bulletin* 21 (1995): 480–90. See also Kim Mooney

and Erica Lorenz, "The Effects of Food and Gender on Interpersonal Perceptions," *Sex Roles* 36 (1997): 639–53.

36. Jeffrey Pilcher, *Qué Vivan Los Tamales!* (Albuquerque: University of New Mexico Press, 1998); David Lind and Elizabeth Barham, "The Social Life of the Tortilla," *Agriculture and Human Values* 21 (2004): 47–60.

37. Harvey Levenstein, *Revolution at the Table* (New York: Oxford University Press, 1988), chap. 8.

38. Levenstein, *Revolution at the Table*, pp. 103–4.

Index

About the author

About the book

Insights,
Interviews
& More . . .

Read on

Meet Barry Glassner

Barry Glassner, born in Roanoke, Virginia, is the author of seven books on contemporary social issues. He is a professor of sociology at the University of Southern California. He graduated from Northwestern University, where he majored in journalism and sociology. He pursued a career as a journalist and was an editor for ABC radio news before receiving his doctorate from Washington University in St. Louis. He became chair of the sociology departments at Syracuse University, the University of Connecticut, and the University of Southern California.

Described by the *New York Times* as "a master at the art of dissecting research," Glassner has published research studies in the *American Sociological Review*, *Social Problems*, and *American Journal of Psychiatry*, among other leading journals in the social sciences. His articles and commentaries have appeared in many newspapers, including the *New York Times*, the *Wall Street Journal*, the *Los Angeles Times*, and *The Chronicle of Higher Education*. He is the recipient of several honors, among them an Outstanding Book of the Year Award from *Choice* magazine and a visiting fellowship at Oxford University.

The *New York Times*, commenting on *The Gospel of Food*, wrote, "Glassner exposes the strained interpretations, 'prejudices dressed up as science,' and pure fabrications behind much received wisdom." The *Los Angeles Times* described the

Jennifer Leshnick

book as "pure fun to read," while The *Wall Street Journal* declared, "Glassner's views about food are on the whole so bracing, it's hard not to root for him as he tilts against modern food dogma."

His previous book, *The Culture of Fear: Why Americans Are Afraid of the Wrong Things* (Basic Books), was named a "Best Book of the Year" by Knight-Ridder newspapers and by the *Los Angeles Times Book Review*. *Kirkus Reviews* called the national bestseller "one of the most important sociological books you'll read this year, and certainly the most reassuring."

Glassner featured prominently in Michael Moore's documentary *Bowling for Columbine*. "I spent a full day and evening with Michael Moore and his film crew touring parts of Los Angeles for what would become his Academy Award-winning documentary," he says. "My one regret is that our meal together—ribs from Phillips Bar-B-Que— ended up on the cutting-room floor. Located in a mini-mall in South Central L.A., Phillips serves some of the best ribs west of Kansas City. Moore and I ate and talked in the adjoining parking lot."

Glassner has appeared on numerous television programs, from *The Oprah Winfrey Show* and *Good Morning America* to *Hardball* and *The O'Reilly Factor*, and on news programs such as CNN, CNBC, and MSNBC. In addition, he has appeared on a variety of radio shows, including National Public Radio's *Fresh Air*, *Talk of the Nation*, and *Marketplace*.

His Sunday morning ritual involves a visit to the Hollywood Farmers' Market. There, he and his wife, literary agent Betsy Amster, purchase "fruits, veggies, and flowers."

A member of the Magic Castle (www .magiccastle.com), Glassner loves to watch great magicians perform. "Although I no longer perform myself," he says, "as a child and adolescent, I was an accomplished amateur magician and edited the magazine published by Magical Youth International (www.magicyouth.com)."

He lives in Los Angeles. ❧

> 'I spent a full day and evening with Michael Moore and his film crew touring parts of Los Angeles. . . . My one regret is that our meal together—ribs from Phillips Bar-B-Que— ended up on the cutting-room floor.'

The Anxiety of Appetite
A Conversation with Barry Glassner

The following interview, conducted by Tracie McMillan, first appeared on Salon.com. An online version remains in the Salon archives. Reprinted with permission.

AMERICA'S FOOD ENTHUSIASTS may find it hard to place the name Barry Glassner. He's not a television chef, or a restaurant critic, or a diet guru. Indeed, the University of Southern California sociologist is known primarily for his bestselling 2000 book, *The Culture of Fear*, a dissection of the anxious underpinnings of the American psyche. It's a subject that might seem to have little relevance to the dinner table, but Glassner begs to differ. If his latest book, *The Gospel of Food*, makes one thing plain, it's that few topics generate more worry among Americans than our breakfasts, lunches, and dinners.

Glassner relishes debate, and *Gospel* takes on nearly every sacred cow of contemporary food culture. High-end restaurant reviewers, eaters seeking "authentic" ethnic eateries, organic converts, local agriculture proponents, and fast food's detractors all receive a methodical interrogation of the accuracy of their claims.

But while Glassner examines nearly every issue populating the food landscape, *Gospel* shines brightest when he turns his gaze to two that are frequently absent from it: poverty and class. Though he places himself in the company of industrial food's most vocal critics, like Michael Pollan and Eric Schlosser, Glassner sets his sights on them, too, questioning the very journalists, writers and advocates who claim to speak truth to power. The problem, he argues, isn't that the Pollans and Schlossers of the world are *wrong*,

66 While Glassner examines nearly every issue populating the food landscape, *Gospel* shines brightest when he turns his gaze to two that are frequently absent from it: poverty and class. 99

About the book

but that they're not exactly *right*, either. To get to the bottom of something as complicated as America's obsession with fast food—a category of dining, he points out, that offers low-income families a clean, affordable, and convenient meal in ways that anti-industrialists seldom acknowledge—he'd rather engage with the complexity of the matter than reduce it to sound bites.

Glassner manages nonetheless to come up with a few maxims of his own: Restaurant reviews rarely reflect the experience of average diners—because most top-notch food critics, disguises or no, are rapidly recognized by chefs, who heap special treatment upon such visitors. The American family meal is *not* dead—in fact, one of the most oft-cited studies in this vein, which attributed the success of a cohort of National Merit Scholars to eating regular family meals, never existed. And obesity is *not* a simple problem of eating less and exercising more; its prevalence among the poor is likely attributable to bingeing brought on by a periodic scarcity of food—not mere ignorance.

Still, Glassner's laundry list of inaccurate spins should not be taken as a humorless diatribe. *Gospel* is also sprinkled with a passionate eater's enthusiasm for cuisines both street and haute. And the extensive journalistic and academic research around which the book revolves is bookended by discussions of the pleasures of food, ranging from Glassner's own horror at a birthday cake devoid of wheat, sugar, milk, and eggs to his hearty enjoyment of a rib joint in south Los Angeles. The point, says Glassner, is that a mix of American puritanism and health obsession has stripped the pleasure out of many Americans' meals—with little to show for it.

Salon recently caught up with Glassner by phone to discuss why trans fat bans aren't all they're cracked up to be and how your mom's advice about food may be the best you'll ever get. ▶

The Anxiety of Appetite *(continued)*

How did the idea for this book come about?

When I finished my previous book, *The Culture of Fear*, it became obvious to me that I had missed one of the biggest fears in American culture, which is that many Americans are afraid of just about every kind of food for one reason or another. So that's what got me started. And secondly, I myself am a big food enthusiast, I love eating diverse foods in very diverse sorts of places, so this gave me an opportunity to do that—and make it tax deductible.

Was there a specific incident that made you recognize Americans' fear of food?

It's the one I start the book with: I was at this birthday party for a child, and I took a bite of the birthday cake and my tongue stuck to the roof of my mouth. The parents were so proud that they had provided this "healthy" birthday cake, because it didn't have anything in it that would make you want to eat a cake. It didn't have eggs, or milk, or wheat, or butter, of course, and it didn't have any sugar because, of course, that could kill you immediately. I started thinking, "It is bizarre that this is what we've come to," and that was kind of the turning point.

You work hard to debunk studies, pulling out ones that contradict commonly held assumptions, or ones that are based on one or two earlier studies or a misrepresentation of findings. What was the most surprising discrepancy you came across?

I guess I'd pick that National Merit study, which supposedly showed that the common denominator among a cohort of National Merit Scholars was that they all had regular family meals. It's mentioned all over the place, when in fact it never existed. The whole truth about family meals

> ❝ I myself am a big food enthusiast, I love eating diverse foods in very diverse sorts of places, so this gave me an opportunity to do that—and make it tax deductible. ❞

was a surprise; I had thought that the family meal was a thing of the past, when in fact families eat together at about the same level as the 1950s.

Food and obesity are such hot topics right now, but you take aim at some very well-respected writers on the subject. For instance, Greg Critser's Fat Land *basically argues that we need to be hardcore about the idea that it's not okay to be fat and, really, lines must be drawn in the sand. But my sense is that you come at the problem from a different angle.*

I think there's a lot of merit about Critser's book and I'm also critical of parts of it. So when he says that the rich are more insightful about what to eat, that concerns me. I think that if we want to understand why it is that people of lower income are more likely to be overweight or obese, to simplify it to a moral condemnation is not a wise way to go. The very notion that wealthy Americans are constrained in their consumption patterns is absurd. Witness their SUVs, their oversize homes. It's fashionable for the wealthy to be thin and eat particular sorts of foods that are on the approved list, but let's not give them undue credit for that, especially at a time when you can go to most any high-end restaurant and get very high-fat, high-calorie meals.

If you want to understand why people of low income tend to be more overweight and obese, it's a complicated story. But we shouldn't leave out the effect that food insecurity itself has; in the book I go into this in some detail, but basically there's a parallel pattern to binge eating, where people who periodically run low on food resemble people who are on diets. When food stamps run out, or the kids' medical expenses take precedence, or the local food bank shuts down or runs out of food, you're not going to eat a lot. And when food becomes available again, you binge. ▶

> 66 The whole truth about family meals was a surprise; I had thought that the family meal was a thing of the past, when in fact families eat together at about the same level as the 1950s. 99

The Anxiety of Appetite *(continued)*

We know that this pattern, this binge pattern, contributes to overweight and obesity. Yet we've come to have this odd notion that it's what people eat, it's what low-income people eat, rather than what they don't eat, or when they don't eat, or which options are not available to them that explains their weight. And moreover, to the extent that heavy people are stigmatized in this country—as they very widely are these days, including by people who see themselves as liberals or progressives—the more we're heaping on further dangers to their health because we know that discrimination itself is a predictor of ill health.

What do you hope to add to the debate?

What I'm hoping is that my book will really open up people's eyes to thinking about some of these topics in ways they haven't before, and in particular, it will make people more open to greater diversity in their diets. I think the good news about this food-obsessed age of ours is that there's a lot of variety out there that wasn't there before. Many Americans take pleasure in exploring new tastes. But at the same time, vast segments of the population greatly restrict what they eat, whether it's because they're on a low-fat diet, a low-carb diet, they shun places that are too popular, or only go places that are very popular. For a whole range of reasons that I write about, they're very restricted, so I think there's a lot more opening up to be done than has happened so far.

What's your sense of how many people are really on these diets? I certainly don't know them—but then, I'm in Brooklyn.

Let me put it this way; I think that the dietary regimens people put themselves on vary and contrast with one another but in many cases are

> 66 To the extent that heavy people are stigmatized in this country . . . the more we're heaping on further dangers to their health because we know that discrimination itself is a predictor of ill health. 99

very restrictive, whether it's veganism or the Atkins diet. And part of what I find interesting is that in each of these cases, followers believe that their diet is miraculous, almost.

My own view is eat and let eat. I'm perfectly comfortable with people following an Atkins diet and eating meat with every meal, or a vegan diet and never eating any animal products. What I'm uncomfortable with are the exaggerated claims that they make, that a meatless regimen can prevent most every serious malady from heart disease to world hunger, or that following an Atkins diet is a magical potion for longevity and weight loss.

I think there are millions and millions of Americans who try to follow one version or another of the "gospel of naught," which is this notion that the worth of a meal lies primarily in what it lacks rather than what it has. So the less sugar, salt, fat, calories, preservatives, animal products, carbs, additives, or whatever the person is concerned about, the better the food. And this seems to me a quite curious notion that's worth a lot more attention than we've given it.

While I was reading the book, I couldn't help feeling a little overwhelmed by all the ways in which commonly held beliefs about what's healthy—whole grains, low fat, being lean—were being called into question. And the big question I came away with was: What do I do now?

I think that there is a basic precept that serves very well, and that's to eat well and enjoyably and moderately over the long haul. I certainly do not advocate that eating a large quantity of any of the substances that are currently considered bad or unhealthy would be a good idea. The kind of diet that Morgan Spurlock went on in *Super Size Me* is obviously going to make you sick. But so would ▶

> **❝** I think there are millions and millions of Americans who try to follow one version or another of the 'gospel of naught,' which is this notion that the worth of a meal lies primarily in what it lacks rather than what it has. **❞**

eating three meals a day of boiled broccoli. So, I think that it's certainly wise to be concerned with eating well and eating moderately and taking into account the sorts of advice that generations of mothers have given, and occasionally fathers. Eat your veggies, eat your fruit, and don't overdose on sweets. So, in no way am I advocating that we should replace the gospel of naught with some kind of absurd diet that would go in the opposite direction.

What about people who say, "Oh, I'm going to treat myself," and end up treating themselves almost daily. How good are Americans at eating moderately?

I think one way that the food industry is brilliant is in picking up on the bipolar approach to food that we have in this country where we think that certain foods are good or bad, or sacred or profane. The food industry will sell us foods that make us feel like we've been good and righteous and then they'll say, often in so many words, "Now that you have been good you can be bad and buy this other product." And they win both ways.

When you listen to a lot of people talk about their meals, they use words like, "I've been bad," if they order a creamy dessert at a meal. Or, "I've been good," if they stay on their diet. The key motivator there is guilt and the avoidance of guilt. And it applies not only to ourselves, but to other people. So many Americans take as a literal truth the old maxim that you are what you eat. We believe that we can tell a lot about a person by what he or she eats when really what we're expressing are prejudices.

In the book I talk about one of my favorite studies, which was a study where students were shown photographs of people their age and researchers told one set of students that the people in the photographs ate foods like whole

> 66 When you listen to a lot of people talk about their meals, they use words like, 'I've been bad,' if they order a creamy dessert at a meal. Or, 'I've been good,' if they stay on their diet. The key motivator there is guilt and the avoidance of guilt. 99

wheat breads and chicken, and they told another set of students that these same people ate hamburgers and French fries and hot fudge sundaes. And in fact, the students had been shown the same people, but they ranked them very differently based on what foods they'd been told they ate; ranked them as more or less likable, more or less attractive. I think that really goes to a deeply ingrained prejudice in society.

Where does that prejudice come from?

I think it comes from our religious backgrounds, which we've taken in this secular direction. In both Judeo and Christian traditions, diet is emphasized and special foods are emphasized. In traditional religious teachings it's very specific; Judaism and Islam prohibited pork, Catholicism decreed fish on Friday. Today we have more secular versions of it, and the response that comes out in an experiment like that is an example of it. People engaging in elaborate rituals to prepare meals; that's another part of it, as is the kind of godlike status of celebrity chefs. I take the titles of my books very seriously, and it was after a lot of thought that I decided to call it *The Gospel of Food*.

I was interested in how you classified different groups of advocates and writers as "food adventurers," those who seek out authentic foods, or "adherents to the gospel of naught," who seek health. How would you classify yourself, and what are you seeking?

I'm very much a food adventurer and a little bit of a foodie. I love expanding my horizons and exploring different cuisines. I'm truly delighted when I'm enjoying ribs from Phillips Bar-B-Que here in south L.A.—and at the same time, I've had some of the most wonderful moments of my life at meals at some of the top restaurants in the ▶

> " I'm very much a food adventurer and a little bit of a foodie. "

The Anxiety of Appetite *(continued)*

country, at places like French Laundry and Daniel. So, I'm kind of odd in that regard.

Food adventurers can be annoyingly focused on finding undiscovered places, but aren't most just saying, "Go out, and eat foods that aren't necessarily in the mainstream"? What's wrong with that?

I'm very enthusiastic about the food adventurers but, as with every other group I discuss in the book, you get a substantial subpopulation of them who just go to extremes or limit what they consider acceptable. In their case, the limits often revolve around the notion of authenticity, and that's a very difficult notion and in the end not terribly helpful. I am critical of food adventurers who dismiss out of hand mainstream reviewers of the sorts of places that they go, who think that if the reviewer for the *Village Voice* or the *L.A. Weekly* discovered a place and liked it, why would you go there? Their point is to go only to places people either wouldn't know about or wouldn't like. When, in fact, there are some great food critics out there, who specialize in the sorts of places food adventurers like to say they discovered.

You take to task one of the more popular ideas among socially conscious foodies like Michael Pollan, which is that America's food is artificially cheap and we should be paying more to help make it possible for workers to earn better wages.

In my mind, you're raising a couple of different issues. It became evident to me in doing this research that the official dietary guidelines are just that. They're guidelines for most of the population. But they're not for the poor who *have* to rely on programs that are required to comply with them. Programs like school meal programs and government-run hospitals, and Women,

Infants and Children are required to comply with the Department of Agriculture's dietary guidelines—and that's one example of many I uncovered in the present and the past, where the poorer you are, the more likely it is you will have food regulations imposed upon you. That's not to say these regulations are bad overall, but we should look at who's affected and in which ways. We should do that as well when we criticize eateries that provide low-cost meals to people on limited income. The fast-food industry deserves a lot of criticism and I level it in the book, but at the same time, to be able to get a complete or nearly complete meal for a few bucks, with distractions for the children thrown in at no extra cost, is not in itself a bad thing. And until those of us on the progressive side of the political spectrum have real alternatives in place, we'd be well advised to look at the good as well as the bad.

I'm certainly critical of the political right for its opposition to minimum wage and various labor laws, but when I see the left focusing so heavily on symbols, rather than on actual conditions, it concerns me. I see relatively little organized attention to hunger, for example, relative to, for instance, the kind of effective and organized campaigns against particular types of foods, like trans fats. When somewhere around thirty-five million to forty million Americans are facing hunger every year it seems to me that *that* would be the top priority of any reasonable food activist. The ban on trans fats may be a good thing, but should it be the first thing? Should it take precedence over much more pressing food issues like hunger in the city, or the availability of fresh foods to the poor in the city? No, not for one minute.

> " The fast-food industry deserves a lot of criticism . . . but at the same time, to be able to get a complete or nearly complete meal for a few bucks, with distractions for the children thrown in at no extra cost, is not in itself a bad thing. "

A One-Word Response
Hunger

WHAT SHOULD BE our biggest concern about our food supply?"

Michael Pollan, Marion Nestle, and I were asked that question by the moderator of a panel on which we spoke at a book festival several months after the hardcover edition of *The Gospel of Food* was published.

Our unrehearsed replies leave no doubt about where I part company with a prevailing assumption among some prominent food writers.

I admire Pollan and Nestle and have learned a great deal from their books and articles, as I have indicated in *The Gospel of Food* and elsewhere. But their replies and mine could not have been more different.

"Calories. In a word, calories," Nestle replied. The biggest public health problem in America today is overweight and the risks it raises for chronic diseases, she said. "We live in a food system that is set up to get us to eat more food rather than less. The entire food system, its entire purpose is to try to get every one of us to eat as much food as we possibly can or spend as much money on it as we possibly can."

Pollan echoed her sentiment and enlarged upon it. "Problems of abundance, manipulation, and confusion," he said, and he elaborated on what he meant by the last of those. "Most eaters have lost track of the fact that when they eat, they are supporting one form of agriculture or another, one sort of relationship with other species or another."

When it was my turn to speak, I began with a one-word response: Hunger.

"We've got millions of Americans who go hungry for part of every month. Many people every night are hungry in America. And there are things we should be doing about this," I put forward. By way of example, I encouraged the audience to write

> 'We've got millions of Americans who go hungry for part of every month. Many people every night are hungry in America. And there are things we should be doing about this.'

to their elected representatives in support of legislation to close the gap on food assistance to children. (According to government statistics, about twenty million children receive food assistance during the school year, but only about two million have access to this help in summer.)

Abundance and excess calories may be problems for some, but should they be "our biggest concern about our food supply"?

I don't think so, not at a time when the World Health Organization reports that one-third of the planet's population is starving, several million people die of hunger each year, and in the United States, the Department of Agriculture estimates that more than eight million Americans experience hunger. Many more suffer from food insecurity. Over the course of a year, twenty-five million Americans receive emergency food assistance from the nation's food banks.

Until those of us who are well fed face up squarely to these realities and find ways to help the hungry, I humbly suggest some humility about matters of concern to the privileged. And for heaven's sake, can we please stop preaching to the poor about their diets?

A week prior to the panel discussion, I received one of the most uplifting letters I can recall from a reader. It came from Susan Douglas, the executive director of a hunger relief organization in South Carolina who had just finished reading *The Gospel of Food*.

"I particularly appreciated your point of view on how we react to low-income families for whom a meal at a fast-food place is a special treat," she wrote. "You also address the imposition of middle class values of the worth of food on programs like ours that work to bring some quality perishables into the diets of those who often turn to pantries for emergency food assistance.

"Each Thanksgiving our organization coordinates a volunteer effort that provides a traditional Thanksgiving meal to around two thousand people. We used to partner with a ▶

A One-Word Response *(continued)*

local outlet of a large chain restaurant and lost the partner this past year. My Board wanted to serve the meals at some of the churches that pitched in to help us out. I pointed out to the Board that for the families and individuals who used to 'dine in' on Thanksgiving, it was not about the meal, it was about going to a place they could not ordinarily afford, sitting down like everyone else, being served by friendly young people and enjoying a special family experience together. We could have served just about anything in the way of food, though we do work hard to make the meal one that would look familiar to any American."

And a great meal it is, for all its calories, carbs, and culinary incorrectness. Would that every American could afford it. ∾